Martin Fuchs

**Topics in the
Calculus of Variations**

Advanced Lectures
in Mathematics

Editorial board:
Prof. Dr. Martin Aigner, Freie Universität Berlin, Germany
Prof. Dr. Gerd Fischer, Heinrich-Heine-Universität Düsseldorf, Germany
Prof. Dr. Michael Grüter, Universität des Saarlandes, Saarbrücken, Germany
Prof. Dr. Manfred Knebusch, Universität Regensburg, Germany
Prof. Dr. Gisbert Wüstholz, ETH Zürich, Switzerland

Martin Fuchs
Topics in the Calculus of Variations

Wolfgang Ebeling
Lattices and Codes

Jesús M. Ruiz
The Basic Theory of Power Series

Heinrich von Weizsäcker and Gerhard Winkler
Stochastic Integrals

Francesco Guaraldo, Patrizia Macri, and Alessandro Tancredi
Topics on Real Analytic Spaces

Manfred Denker
Asymptotic Distribution Theory in Nonparametric Statistics

Jochen Werner
Optimization. Theory and Applications

Martin Fuchs

Topics in the Calculus of Variations

Prof. Dr. Martin Fuchs
Fachbereich Mathematik
Universität des Saarlandes
Postfach 15 11 50
66041 Saarbrücken

Mathematics Subject Classification:
49 N 60, 49 Q 15, 49 Q 20, 58 E 20, 58 E 30, 58 E 35, 73 C 50

All rights reserved
© Friedr. Vieweg & Sohn Verlagsgesellschaft mbH, Braunschweig/Wiesbaden, 1994

Vieweg is a subsidiary company of Bertelsmann Professional Information.

No part of this publication may be reproduced, stored in a retrieval system or transmitted, mechanical, photocopying or otherwise, without prior permission of the copyright holder.

Cover design: Klaus Birk, Wiesbaden
Printing and binding: Lengericher Handelsdruckerei, Lengerich
Printed on acid-free paper
Printed in Germany

ISSN 0932-7134
ISBN 3-528-06623-7

Preface

This report grew out of a series of lectures given at the East China Institute of Technology, Nanjing, during September 1992. The purpose of this book is to make beginning research students familiar with some problems in variational calculus which have been chosen following my personal taste but with the attempt to illustrate two basic principles in the calculus of variations which are the fundamental question of existence of (generalized) solutions and closely related the question of regularity. Chapter one is devoted to the study of variational integrals for vectorvalued functions which began with the pioneering work of Morrey [68] in the thirties. We concentrate on problems where also nonlinear side conditions are imposed on the classes of admissible comparison functions. As special cases we include mappings whose range is forced to lie in some Riemannian manifold possibly with boundary or functions whose Jacobian is required to be strictly positive. The variational integrals under consideration are typically nonquadratic with respect to the gradient which immediately leads us to classes of degenerate elliptic systems. Let us mention some of the most important applications:

- p–harmonic maps between Riemannian manifolds

- systems of degenerate variational inequalities

- model problems in nonlinear elasticity.

Usually by working in appropriate Sobolev spaces, the existence of generalized solutions is rather easy to establish (see [7]) but leads to apriori discontinuous functions. In the vectorial case partial regularity is a wellknown phenomenon (compare [22]) and we present a collection of methods which show that even in the presence of nonlinear constraints one can expect differentiability of solutions up to a closed set of small measure. Of course, the size of the present book is limited which makes it impossible to discuss all the proofs in great detail. So we made a selection concentrating on the basic theorems A–F from section 1.1, the proofs of all other results are briefly indicated and can be traced in the literature.

In chapter two we treat a purely geometric problem by trying to generalize the concept of surfaces of prescribed mean curvature in Euclidean three-space to the case of arbitrary dimension and codimension. Although this topic seems to be very special it should be seen as a guideline of how to formulate and to prove existence for geometric variational problems in the setting of Geometric Measure Theory. We have included some background material on Geometric Measure Theory since we feel that the reader should get familiar with this important field. Concerning the regularity of generalized manifolds with prescribed mean curvature form we have no essential contribution apart from the codimension one case for which the same results as for area minimizes are true.

Although chapters one and two are devoted to formally different subjects there are strong connections which should be pointet out: first, arguments like monotonicity, blow-up and tangent behaviour originated in Geometric Measure Theory and are nowadays standard in the partial regularity theory of vectorvalued minimizers. Second, the method of linearisation applies to obstacle problems for mappings as well as to the existence theory of generalized mean curvature manifolds. Third, the isoperimetric inequality is used for example to prove regularity for approximations in nonlinear twodimensional elasticity. On the other hand Almgren's optimal isoperimetric inequality serves as a basis for our investigations in chapter two.

I would like to thank Professor Yang Xiaoping for inviting me to visit the East China Institute of Technology, and I appreciate the friendly atmosphere provided by the members of the Mathematical Department. Thanks are also due to Mrs. M. Tabbert for doing an excellent typing job. I also acknowledge financial support from the Deutsche Forschungsgemeinschaft.

Saarbrücken, June 1994   Martin Fuchs

Contents

1 Degenerate Variational Integrals with Nonlinear Side Conditions, p–harmonic Maps and Related Topics   1
    1.1 Introduction, Notations and Results for Minimizers   1
    1.2 Linearisation of the Minimum Property, Extension of Maps .   8
    1.3 Proofs of the Basic Theorems   22
    1.4 A Survey on p-Harmonic Maps   40
    1.5 Variational Inequalities and Asymptotically Regular Integrands   58
    1.6 Approximations for some Model Problems in
         Nonlinear Elasticity.   74

2 Manifolds of Prescribed Mean Curvature in the Setting of Geometric Measure Theory   84
    2.1 The Mean Curvature Problem   84
    2.2 Some Facts from Geometric Measure Theory   97
    2.3 A First Approach to the Mean Curvature Problem 109
    2.4 General Existence Theorems, Applications to Isoperimetric
         Problems 119
    2.5 Tangent Cones, Small Solutions, Closed Hypersurfaces 128

Bibliography   139

Index   145

1 Degenerate Variational Integrals with Nonlinear Side Conditions, p–harmonic Maps and Related Topics

1.1 Introduction, Notations and Results for Minimizers

We are concerned with the (partial) regularity properties of vectorvalued Sobolev functions $u : \mathbb{R}^n \supset \Omega \to \mathbb{R}^N$ which minimize certain degenerate variational integrals $\mathcal{E}(u, \Omega) := \int_\Omega f(\cdot, u, \nabla u)dx$ subject to some nonlinear side conditions imposed on the admissible comparison functions, i.e. we consider vectorvalued obstacle problems for functionals with nonstandard growth. Before giving some historical comments we fix our notations and state the results.

In the *Euclidean case* Ω is a bounded subdomain of $\mathbb{R}^n$, $n \geq 2$, and $M \subset \mathbb{R}^N$ denotes a smooth domain with $\partial M \in C^2$. For $p \in (1, \infty)$

$$H^{1,p}(\Omega, \overline{M}) := \{u \in H^{1,p}(\Omega, \mathbb{R}^N) : u(x) \in \overline{M} \; \mathcal{L}^n - \text{a.e.}\}$$

is the restricted class of Sobolev functions with values in the closure $\overline{M}$ of M.

Suppose further that we are given functions

$$a_{\alpha\beta} = a_{\beta\alpha} : \overline{\Omega} \to \mathbb{R}, \quad B^{ij} = B^{ji} : \overline{\Omega} \times \mathbb{R}^N \to \mathbb{R}$$
$$\alpha, \beta = 1, \ldots, n, \quad i, j = 1, \ldots, N$$

of class C^1 satisfying the ellipticity estimates

$$\begin{cases} a_{\alpha\beta}(x) \eta_\alpha \eta_\beta \geq \mu \cdot |\eta|^2 \\ B^{ij}(x, y) \zeta^i \zeta^j \geq \mu \cdot |\zeta|^2 \end{cases}, \; (x, y) \in \overline{\Omega} \times \mathbb{R}^N, \; \eta \in \mathbb{R}^n, \; \zeta \in \mathbb{R}^N$$

for some positive μ. We let

$$\mathcal{F}(u, \Omega) := \int_\Omega \left(a_{\alpha\beta}(x) B^{ij}(x, u) \partial_\alpha u^i \partial_\beta u^j\right)^{p/2} dx.$$

Finally, we define the regular and singular set of an arbitrary function $u \in H^{1,p}(\Omega, \mathbb{R}^N)$:

$\text{Reg}(u) := \{x \in \Omega : \ u \text{ is continuous in a neighborhood of } x\}$

(interior regular set)

$\text{Sing}(u) := \Omega - \text{Reg}(u)$

(interior singular set)

Note that $\text{Reg}(u) = \Omega$ for $p > n$ by Sobolev's embedding theorem but in the general case $1 \leq p \leq n$ no apriori information on $\text{Reg}(u)$ is available. By definition $\text{Reg}(u)$ is open but the regular set may be empty. A function $u \in H^{1,p}(\Omega, \overline{M})$ is said to be locally $\mathcal{F}$-minimizing in $H^{1,p}(\Omega, \overline{M})$ iff

$$\mathcal{F}(u, G) \leq \mathcal{F}(v, G)$$

holds for any open subregion G of Ω and any $v \in H^{1,p}(\Omega, \overline{M})$ such that $\text{spt}\,(u - v) \subset\subset G$.

In the *Riemannian case* Ω denotes an open subregion of some n–dimensional Riemannian manifold, Y is a k–dimensional submanifold of Euclidean space $\mathbb{R}^N$. We let

$$M := \begin{cases} \text{open subregion of } Y \text{ with } \emptyset \neq \partial M \in C^2 \\ \text{and } \overline{M} \text{ compactly contained in } \text{Int}(Y) \\ \text{or} \\ Y \text{ in case } Y \text{ is compact with } \partial Y = \emptyset \,. \end{cases}$$

In the first case we have a Riemannian obstacle problem (e.g. $Y = S^{N-1}$ and $M = \{y \in Y : y^N > 0\}$), in the second case we just consider mappings with values in some submanifold of $\mathbb{R}^N$ having no boundary. Clearly this situation is much easier and follows as a corollary from our general discussion on obstacle problems.

With the help of local coordinates in Ω we define the spaces

$$H^{1,p}(\Omega, \mathbb{R}^N) \supset H^{1,p}(\Omega, Y) \supset H^{1,p}(\Omega, \overline{M})$$

and the energy

$$\mathcal{E}(u, \Omega) := \int_\Omega \|du\|^p \, d\,\text{vol}$$

for functions in these spaces. The notions of locally minimizing maps and of $\text{Reg}(u)$, $\text{Sing}(u)$ are as above.

Theorem A (First estimate on $\text{Sing}(u)$): *Suppose that $u \in H^{1,p}(\Omega, \overline{M})$ is locally $\mathcal{F}$-minimizing. Then*

1.1 Introduction, Notations and Results for Minimizers

$$\operatorname{Sing}(u) = \{x \in \Omega : \liminf_{r \searrow 0} r^{p-n} \int_{B_r(x)} |\nabla u|^p \, dz > 0\},$$

especially $\mathcal{H}^{n-p}(\operatorname{Sing} u) = 0$ *and* $\operatorname{Sing}(u) = \emptyset$ *for* $p \geq n$.
The same result holds in the Riemannian case.

Theorem B (Optimal interior partial regularity): *Under the assumptions of A we have*

(i) $\mathcal{H} - \dim(\operatorname{Sing}(u)) \leq n - [p] - 1$ *if* $n > p+1$

  ($[p] := \max\{\ell \in \mathbb{N} : \ell \leq p\}$)

(ii) $\operatorname{Sing}(u)$ *is discrete (= no interior accumulation points) for* $n-1 \leq p < n$.

Here $\mathcal{H}^\ell$ denotes ℓ-dimensional Hausdorff measure and

$$\begin{cases} \mathcal{H} - \dim(A) = h & :\Leftrightarrow \\ \mathcal{H}^{h+\varepsilon}(A) = 0 & \text{for all } \varepsilon > 0, \\ \mathcal{H}^h(A) > 0 \end{cases}$$

is the definition of Hausdorff's dimension. Note that in a measure theoretic sense Theorem B says that the dimension of $\operatorname{Sing}(u)$ can be reduced "from $n-p$ to $n-p-1$".

Up to now nothing is known about the analytic structure of $\operatorname{Sing}(u)$ (pieces of manifolds in Ω?) so that the word dimension has no actual geometric meaning.

Theorem C: *If u is locally $\mathcal{F}$- or $\mathcal{E}$-minimizing in the restricted class then $u \in C^{1,\alpha}(\operatorname{Reg} u)$ for some $\alpha \in (0,1)$.*

Even for scalar variational inequalities with analytic data and $p = 2$ C^2-regularity is unnatural. The best one can expect is Lipschitz continuity of ∇u. But even for free local minimizers of $\int_\Omega |\nabla u|^p \, dx$ Giaquinta & Modica [47] gave an example which is only of class $C^{1,\varepsilon}$ for some small $\varepsilon > 0$. So the lack of higher regularity is not only due to the presence of the side condition but is also caused by the degeneracy of the functional.

Theorem D (Behaviour at isolated singularities): *Let $p \in [n-1, n)$ and suppose that $u \in H^{1,p}(\Omega, \overline{M})$ is locally minimizing. Then if $x_0 \in \operatorname{Sing}(u)$ we have*

$$\limsup_{x \to x_0} |\nabla u(x)| \cdot \operatorname{dist}(x, x_0) < \infty.$$

Theorem E (Boundary regularity): *Suppose that $\partial\Omega$ is smooth and let $u_0 \in H^{1,p}(\Omega, \overline{M})$ denote a function which is smooth near and on $\partial\Omega$. Then if $u \in H^{1,p}(\Omega, \overline{M})$ is a minimizer for boundary values u_0, u is Hölder continuous on $\{x \in \overline{\Omega} : \mathrm{dist}(x, \partial\Omega) \leq \rho\}$ for some $\rho > 0$.*

In the simple case
$$\begin{cases} n = 3 = N, & p = 2 \\ \Omega, M \text{ domains in } \mathbb{R}^3 \end{cases}$$
any minimizer u of $\int_\Omega |\nabla u|^2 \, dx$ in $H^{1,2}(\Omega, \overline{M})$ for boundary values $u_0 \in H^{1,2}(\Omega, \overline{M})$ which are smooth near $\partial\Omega$ behaves in the following way:

> There is a finite number $x_1, \ldots, x_L$ of interior points of Ω such that $u \in C^{0,\alpha}(\overline{\Omega} - \{x_1, \ldots, x_L\})$.
> Moreover, $u \in C^{1,\beta}(\Omega - \{x_1, \ldots, x_L\})$ and near each possible singular point x_i we have $u(x) \approx (x - x_i)/|x - x_i|$.

Singularites of minimizers may occur for two reasons. First, we have *topological obstructions*: Let $n = N = 3$, $p = 2$, $\Omega = B^3$, $M = \{y \in \mathbb{R}^3 : \frac{1}{2} < |y| < 2\}$ and $u_0(x) := |x|^{-1} \cdot x$ which is in the space $H^{1,2}(B^3, \overline{M})$ and smooth near ∂B^3. Then the corresponding minimizer u of $\int_{B^3} |\nabla u|^2 \, dx$ must be discontinuous, i.e. $\# \mathrm{Sing}(u) \geq 1$. It would be interesting to calculate a solution for this configuration. But sometimes $\mathrm{Sing}(u) \neq \emptyset$ is not as obvious as above. In the paper [26] we considered the region

$$\overline{M} := \{z \in \mathbb{R}^{n+1} : 1 \leq |z| \leq 3,\ z^n \leq 0\}$$
$$\cup \{(z', z^{n+1}) \in \mathbb{R}^{n+1} : z^{n+1} \geq 0,\ 1 \leq |z'| \leq 3\}$$

and the boundary map $u_0 : B^n \ni x \mapsto (|x|^{-1} \cdot x, 0)$ for which we proved minimality w.r.t. $\int_{B^n} |\nabla w|^2 \, dx$ in the appropriate class at least for $n \geq 7$. Clearly there are smooth maps $u : B^n \to \overline{M}$ s.t. $u(x) = u_0(x)$ on ∂B^n. In this example it needs less energy for the minimizer to create singular points than to be smooth on $\overline{B^n}$ and staying inside $\overline{M}$.

Finally, we consider sets M for which $\mathrm{Sing}(u) = \emptyset$.

Theorem F (Everywhere regularity): *Suppose that $u \in H^{1,p}(\Omega, \overline{M})$ is a minimizer of $\int_\Omega |\nabla u|^p \, dx$ in the Euclidean case and that there is a point $\zeta \in M$ such that $\zeta + t(z - \zeta) \in M$ for all $0 \leq t < 1$ and $z \in \partial M$. Then*

$\text{Sing}(u) = \emptyset$. *In the Riemannian case we have the same result provided $\overline{M}$ is contained in a regular ball around ζ and the geodesics $\gamma(t)$ from ζ to $z \in \partial M$ stay inside M for all $t < 1$.*

The condition on M excluding singular points roughly states that M is strongly starshaped w.r.t. some interior point ζ. Clearly such configurations are general enough to study contact problems which means that one easily can give examples of sets M and boundary values u_0 for which the minimizer must contact ∂M. One should try to describe the analytic properties of the contact set $[u \in \partial M] = \{x \in \Omega : u(x) \in \partial M\}$ for such configurations, a first unsatisfactory attempt has been made in [28]. Note that in the above counter example of a singular minimizer the set M is a limit of strongly star shaped domains.

Our lectures are organized as follows: In section 1.2 we present some basic tools which are needed to prove the Theorems A – F. The next chapter gives an outline of how to obtain regularity for minimizers where we concentrate on the interior partial regularity theorem A. Section 1.4 studies p–harmonic maps $u : X \to Y$ of Riemannian manifolds X, Y with $\partial Y = \emptyset$ by the way extending wellknown results for harmonic maps ($p = 2$). At the end of chapter 1 we present a collection of results including degenerate systems of variational inequalities and variational integrals occuring in nonlinear elasticity.

But before starting with our program we have a brief look at the history of general partial regularity theory in the vectorvalued case:

I. free minimizers of quadratic variational integrals

II. energy minimizing harmonic maps of Riemannian manifolds

III. quadratic obstacle problems

IV. free minimizers of degenerate variational integrals

V. degenerate variational integrals with nonlinear side conditions.

I. Consider for simplicity the functional

$$\mathcal{F}_2(u, \Omega) := \int_\Omega A^{ij}_{\alpha\beta}(\cdot, u) \partial_\alpha u^i \partial_\beta u^j \, dx$$

with smooth elliptic coefficients $A^{ij}_{\alpha\beta}$. If $u \in H^{1,2}(\Omega, \mathbb{R}^N)$ is a minimizer then Morrey showed in the forties that

$$\text{Sing}(u) = \emptyset \quad \text{if } n = 2 \,.$$

For $n \geq 3$ there are examples of singular minimizers (of the type $\frac{x}{|x|}$) even in the case of splitting functionals. Around 1980 Giaquinta & Giusti [45], [46], compare also Giaquinta's book [44], proved partial regularity in the sense that

$$\mathcal{H}^{n-2}(\text{Sing } u) = 0.$$

The main ingredient is Caccioppoli's inequality

$$\fint_{B_{R/2}} |\nabla u|^2 \, dx \leq c \cdot R^{-2} \fint_{B_R} |u - (u)_R|^2 \, dx \quad (1.1.1)$$

which follows easily from $\mathcal{F}_2(u, B_R) \leq \mathcal{F}_2(u + \eta[(u)_R - u], B_R)$ where $\eta \in C_0^1(B_R, [0, 1])$. Note that for problems with constraints $u + \eta[(u)_R - u]$ is no longer admissible. A second tool is the local comparison with the minimizer of a frozen problem (with constant coefficients) for which good regularity estimates hold. Again, the minimizer of the frozen problem does in general not respect a side condition. Let us remark that (1.1.1) for example implies $\nabla u \in L_{\text{loc}}^{2+\delta}$ for some $\delta > 0$.

II. A related problem occurs when minimizing

$$\mathcal{E}_2(u, \Omega) := \int_\Omega \|du\|^2 \, d\,\text{vol}$$

in the class $H^{1,2}(\Omega, Y)$ where $Y \hookrightarrow \mathbb{R}^N$ is a submanifold of $\mathbb{R}^N$ with $\partial Y = \emptyset$. In local coordinates $\mathcal{E}_2$ takes the form $\mathcal{F}_2$ with splitting coefficients and under the assumption that the range of minimizer u can be covered by a single chart the technique of Giaquinta & Giusti implies $\mathcal{H}^{n-2}(\text{Sing } u) = 0$. Schoen & Uhlenbeck [71], [72] extended $\mathcal{H}^{n-2}(\text{Sing } u) = 0$ to the general case with the help of an elaborate mollification technique. Their proof is restricted to the case $p = 2$ since they use the Euler system $\Delta u = f(\cdot, u, \nabla u)$ combined with the property $\Delta(u^\rho) = (\Delta u)^\rho$ for mollifications.

III. Quadratic obstacle problems wihtout imposing convexity or unnatural smallness assumptions were first attacked by the author [25] and later on in a series of (joint) papers by Duzaar and myself, I refer to the list of references given in [30].

Our proofs do not extend to $p \neq 2$ since we are argued either by mollification (and used $L(u^\rho) = (Lu)^\rho$ for the linear leading part of the Euler operator) or by blow up (using $L(u^i) \to L(u)$ in the sense of distributions if $u^i \rightharpoonup u$ weakly in $H^{1,2}$).

1.1 Introduction, Notations and Results for Minimizers

Remark: Special cases as $n = 2$ or convex "small" M have been treated earlier by several authors, we refer to [62] and the references quoted therein.

IV. The first contribution to the regularity of free local minimizers u : $\mathbb{R}^n \supset \Omega \to \mathbb{R}^N$ of $\int_\Omega |\nabla u|^p \, dx$ with $p \neq 2$ is due to [80] who showed $C^{1,\mu}(\Omega)$–regularity by deriving a differential inequality (via difference quotient technique) for $|\nabla u|^p$ and applying a maximum-principle. Clearly this approach has no extension to problems with constraints. The more general functional $\int_\Omega \left(a_{\alpha\beta}(\cdot, u) B^{ij}(\cdot, u) \partial_\alpha u^i \partial_\beta u^j \right)^{p/2} dx$ was attacked independently by Giaquinta & Modica [47] and Fusco & Hutchinson [43]. They proved $\mathcal{H}^{n-p}(\mathrm{Sing}\, u) = 0$ for free local minimizers in $H^{1,p}(\Omega, \mathbb{R}^N)$ relying on a Caccioppoli type inequality (1.1.1) with p in place of the exponent 2 using the same comparison function. As a second tool they make use of the Uhlenbeck result for a frozen functional. Both steps break down when dealing with an additional side condition.

V. In 1987 Hardt & Lin [53] published a paper in which they claimed $\mathcal{H}^{n-p}(\mathrm{Sing}\, u) = 0$ for minimizers $u \in H^{1,p}(\Omega, Y)$ of $\int_\Omega |du|^p d\,\mathrm{vol}$ where Y is a Riemannian manifold with $\partial Y = \emptyset$. But their proof contains a gap since they use

$$\begin{cases} u_i \rightharpoonup u \text{ in } H^{1,p} \Longrightarrow \\ \partial_\alpha(|\nabla u_i|^{p-2} \partial_\alpha u_i) \longrightarrow \partial_\alpha(|\nabla u|^{p-2} \partial_\alpha u) \\ \text{in the sense of distributions} \end{cases}$$

which is wrong without further information on the sequence under consideration. (One can prove the above statement with the help of the partial regularity theory. Very recently I could close this gap by showing $\partial_\alpha(|\nabla u|^{p-2} \partial_\alpha u) = 0$ under the additional assumption

$$\int_\Omega |\nabla u_i|^{p-2} \nabla u_i \cdot \nabla \varphi \, dx \leq c_i \cdot \|\varphi\|_{L^\infty(\Omega)}$$

for all $\varphi \in \overset{\circ}{H}{}^{1,p} \cap L^\infty(\Omega, \mathbb{R}^N)$ with $\lim\limits_{i \to \infty} c_i = 0$. The proof is rather technical and makes crucial use of arguments due to Frehse and Landes [63].)

My research on p–harmonic obstacle problems started in the year 1985 and the basic theorems A–F were published as Habilitations thesis at Düsseldorf University in 1987. We should also mention a paper of Luckhaus [66] who gave an independent proof of the partial regularity of minimizing p–harmonic maps. Nowadays we have a rather satisfying theory of the subject but nevertheless there remain a lot of hard unsolved problems, for example the extension to nonsplitting functionals $\int_\Omega \left(A^{ij}_{\alpha\beta}(\cdot,u)\partial_\alpha u^i \partial_\beta u^j\right)^{p/2} dx$ even in case $p=2$.

1.2 Linearisation of the Minimum Property, Extension of Maps

If $\Sigma \subset \mathbb{R}^k$ is a compact N–dimensional Riemannian manifold *without boundary* and if $u \in H^{1,p}(\Omega, \mathbb{R}^k)$ minimizes $\int_\Omega |\nabla u|^p dx$ under the side condition $u(x) \in \Sigma$ a.e. in Ω then it is well known how to get the Euler system:

Let π denote the smooth nearest point projection onto Σ defined on a uniform tabular neighborhood of Σ. Then

$$\frac{d}{dt/0}\int_\Omega |\nabla \pi(u+t\Phi)|^p dx = 0$$

for any $\Phi \in C^1_0(\Omega, \mathbb{R}^k)$ (or $\Phi \in \overset{\circ}{H}{}^{1,p}(\Omega,\mathbb{R}^k) \cap L^\infty$)

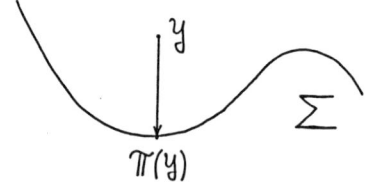

$$\Longrightarrow \quad 0 = \frac{1}{2}p \int_\Omega |\nabla u|^{p-2} \frac{\partial}{\partial t/0}\left[\nabla\pi(u+t\Phi)\cdot\nabla\pi(u+t\Phi)\right]dx$$

$$= p \cdot \int_\Omega |\nabla u|^{p-2} \nabla u \cdot \nabla[D\pi(u)\Phi]dx$$

$$\uparrow$$

derivative of π at u applied to Φ

Look at the expression $\partial_\alpha u \cdot \partial_\alpha[D\pi(u)\Phi]$:

$$\partial_\alpha u \cdot \partial_\alpha[D\pi(u)\Phi] = \partial_\alpha u \cdot D\pi(u)(\partial_\alpha \Phi) + \partial_\alpha u \cdot \partial_\alpha(D\pi(u))\Phi$$

$$= \partial_\alpha u \cdot D\pi(u)(\partial_\alpha \Phi) + \Phi \cdot A_u(\partial_\alpha u, \partial_\alpha u),$$

$A_y(\tau, \eta) := 2^{\text{nd}}$ fundamental form of Σ at y applied to $\tau, \eta \in T_y\Sigma$.

Observe:

$$\partial_\alpha u(x) \in T_{u(x)}\Sigma \quad \text{for a.a. } x \in \Omega. \tag{1.2.1}$$

1.2 Linearisation of the Minimum Property, Extension of Maps

Hence:
$$\begin{aligned}0 &= \int_\Omega |\nabla u|^{p-2}\, \partial_\alpha u \cdot D\pi(u)(\partial_\alpha \Phi)\, dx \\ &\quad + \int_\Omega |\nabla u|^{p-2}\, \Phi \cdot A_u(\partial_\alpha u, \partial_\alpha u)\, dx\end{aligned}$$

The linear map $D\pi(y) : \mathbb{R}^k \to T_y\Sigma$ is just the projection on $T_y\Sigma$ so that
$$\partial_\alpha u \cdot D\pi(u)(\partial_\alpha \Phi) = D\pi(u)(\partial_\alpha u) \cdot \partial_\alpha \Phi = \partial_\alpha u \cdot \partial_\alpha \Phi$$
on account of (1.2.1). Thus we arrive at
$$0 = \int_\Omega |\nabla u|^{p-2}\{\nabla u \cdot \nabla \Phi + \Phi \cdot A_u(\partial_\alpha u, \partial_\alpha u)\}\, dx,$$
i.e.:
$$\begin{cases} u \text{ is a weak solution of} \\ \partial_\alpha(|\nabla u|^{p-2}\, \partial_\alpha u) = |\nabla u|^{p-2}\, A_u(\partial_\alpha u, \partial_\alpha u) \end{cases} \quad (1.2.2)$$

Remarks:

1. We assume that Ω is a flat domain in some space $\mathbb{R}^n$. For Riemannian Ω the situation is essentially the same.

2. For $p = 2$ (1.2.2) reduces to the well known harmonic map equation.

3. Observe that (1.2.2) is degenerate elliptic and that the right-hand-side is of growth order p in ∇u ($\to$ critical growth because we work in the space $H^{1,p}$).

We now pass to derive an Euler system for the case of obstacle problems. The idea of linearisation goes back to [25] and later on was refined by Duzaar and myself in several papers, we refer for example to [8]. The present version is taken from [30].

Suppose $\Omega \subset \mathbb{R}^n$ is bounded and open and that $M \subset \mathbb{R}^N$ is a bounded domain with ∂M of class C^2. Let $u \in H^{1,p}(\Omega, \mathbb{R}^N)$ denote a constrained local minimizer of $\int_\Omega |\nabla u|^p dx$ subject to the side condition $u(x) \in \overline{M}$ a.e.

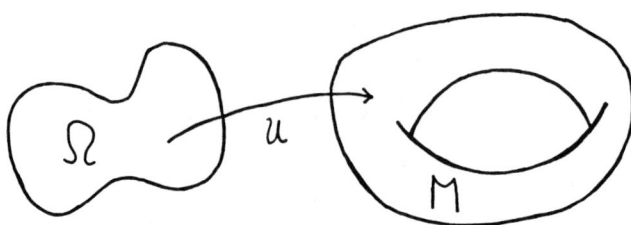

In contrast to our previous considerations two different cases may occur (and are actually observed):

$$u(x) \in \text{Interior of } M \quad \text{or}$$
$$u(x) \in \Sigma = \partial M \quad (N-1 \text{ dim. manifold})$$

In the first case one could expect the equation

$$\partial_\alpha(|\nabla u|^{p-2}\partial_\alpha u) = 0$$

to hold, in the second case equation (1.2.2). By combining the two cases one is intuitively let to the conjecture that now the local minimzer is a weak solution of a system of the form

$$\partial_\alpha(|\nabla u|^{p-2}\partial_\alpha u) = \mathbf{1}_{[u\in\Sigma]} \cdot |\nabla u|^{p-2} A_u(\partial_\alpha u, \partial_\alpha u) \qquad (1.2.3)$$

on Ω where $[u \in \Sigma] := \{x \in \Omega : u(x) \in \Sigma\}$ and $\mathbf{1}_{[\ldots]}$ is the characteristic function of this set coupling the two different cases.

In order to make things precise we introduce some

Notations:
Since $\partial M = \Sigma$ is a C^2-hypersurface the function

$$d(z) := \text{dist}\,(z, \Sigma)$$

is smooth for $z \in \overline{M}$ near Σ. By *negative reflection* we extend d to a uniform tubular neighborhood $\mathcal{U}$ of Σ, finally let

$$\mathcal{N} := \text{grad}\, d \quad \text{on } \mathcal{U}.$$

1.2 Linearisation of the Minimum Property, Extension of Maps

Clearly $\mathcal{N}(z)$ is just the interior unit normal to ∂M for points $z \in \partial M$.

In order to obtain the Euler equation there are two kinds of admissible variations: push u along $\mathcal{N}$ into the interior of M or use tangential variations.

Normal variations:
For small $0 < \varepsilon < 1$ consider
$$h_\varepsilon : [0, \infty) \to [0, 1]$$
smooth such that
$$h_\varepsilon(s) = \begin{cases} 1 &, \quad s \leq \varepsilon \\ 0 &, \quad s \geq 2\varepsilon \end{cases}, \quad h'_\varepsilon \leq 0,$$
for example

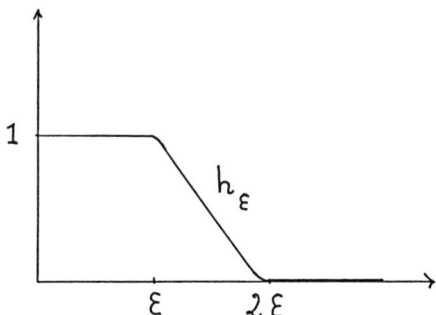

and define
$$u_t := u + t \cdot \eta \cdot h_\varepsilon(du) \cdot \mathcal{N}(u)$$
with $\eta \in C_0^1(\Omega)$, $\eta \geq 0$, $0 \leq t << 1$. (If $u(x)$ is near Σ then $h_\varepsilon(du) = 1$ and because of the sign properties of t and η $u(x)$ is pushed into the interior of M. In case that $d(u(x))$ is "large" $u_t(x)$ is just $u(x)$.)

From $u_t(x) \in \overline{M}$ and $\operatorname{spt}(u_t - u) \subset \operatorname{spt} \eta \subset\subset \Omega$ we deduce
$$\lim_{t \downarrow 0} \frac{1}{t} \cdot \left\{ \int_\Omega |\nabla u_t|^p dx - \int_\Omega |\nabla u|^p dx \right\} \geq 0$$
and by Riesz representation there is a nonnegative Radon measure λ such that
$$\int_\Omega p|\nabla u|^{p-2} \partial_\alpha u \cdot \partial_\alpha(\eta \cdot h_\varepsilon(du)\mathcal{N}(u)) \, dx = \int_\Omega \eta \, d\lambda, \quad \forall \eta \in C_0^1(\Omega). \quad (1.2.4)$$

The measure λ is independent of ε: for $\varepsilon \neq \varepsilon'$ it is easy to see that the variation

$$v_t := u + t \cdot \varphi \cdot [h_\varepsilon(d(u)) - h_{\varepsilon'}(d(u))] \mathcal{N}(u)$$

is admissible for all $\varphi \in C_0^1(\Omega)$ and $|t| << 1$ which implies

$$\int_\Omega p \cdot |\nabla u|^{p-2} \partial_\alpha u \cdot \partial_\alpha \Big(\varphi[h_\varepsilon(d(u)) - h_{\varepsilon'}(d(u))] \mathcal{N}(u)\Big) dx = 0$$

and in conclusion $\lambda_\varepsilon = \lambda_{\varepsilon'}$. In order to obtain information on λ we want to pass to the limit $\varepsilon \downarrow 0$ in (1.2.4). Breaking up $\partial_\alpha(\eta \cdot h_\varepsilon(du) \mathcal{N}(u))$ the left-hand-side of (1.2.4) splits into three terms for which we get: ($\eta \geq 0$)

$$\int_\Omega p|\nabla u|^{p-2} \partial_\alpha u \cdot \partial_\alpha \eta \, h_\varepsilon(d(u)) \mathcal{N}(u) \, dx \xrightarrow[\varepsilon \downarrow 0]{} \int_{[u \in \Sigma]} p|\nabla u|^{p-2} \partial_\alpha u \, \partial_\alpha \eta \cdot \mathcal{N}(u) \, dx = 0$$

on account of $\partial_\alpha u \cdot \mathcal{N}(u) = \partial_\alpha(d \circ u) = 0$ a.e. on the set $[u \in \Sigma] = [d(u) = 0]$,

$$\int_\Omega p \cdot |\nabla u|^{p-2} \partial_\alpha u \cdot \eta \, \partial_\alpha \Big(h_\varepsilon(d(u))\Big) \mathcal{N}(u) \, dx$$

$$= \int_\Omega p|\nabla u|^{p-2} \partial_\alpha u \cdot \eta \cdot h'_\varepsilon(d(u)) \, (\mathcal{N}(u) \cdot \partial_\alpha u) \cdot \mathcal{N}(u) \, dx \leq 0$$

(recall $\eta \geq 0$, $h'_\varepsilon \leq 0$),

$$\int_\Omega p \cdot |\nabla u|^{p-2} \partial_\alpha u \cdot \eta \cdot h_\varepsilon(du) \, \partial_\alpha(\mathcal{N} \circ u) dx$$

$$\xrightarrow[\varepsilon \downarrow 0]{} \int_{[u \in \Sigma]} p \cdot |\nabla u|^{p-2} \eta \cdot \partial_\alpha u \cdot \partial_\alpha(\mathcal{N} \circ u) \, dx$$

$$= \int_{[u \in \Sigma]} \eta \cdot p|\nabla u|^{p-2} A_u(\partial_\alpha u, \partial_\alpha u) \, dx \, ,$$

where A_u is the second fundamental form of $\Sigma = \partial M$. Collecting our results we have shown:

$$\int_\Omega \eta \, d\lambda \leq \int_\Omega \mathbf{1}_{[u \in \Sigma]} \, p \cdot |\nabla u|^{p-2} A_u(\partial_\alpha u, \partial_\alpha u) \eta \, dx$$

for any $\eta \in C_0^1(\Omega)$, $\eta \geq 0$, especially

$$A_u(\partial_\alpha u, \partial_\alpha u) \geq 0 \quad \text{a.e. on } [u \in \Sigma] \, . \tag{1.2.5}$$

The Radon Nikodym theorem gives the existence of a density function $\Theta : \Omega \to [0, 1]$ such that

1.2 Linearisation of the Minimum Property, Extension of Maps

$$\int_\Omega \eta\, d\lambda = \int_\Omega \Theta \cdot p \cdot |\nabla u|^{p-2} A_u(\partial_\alpha u, \partial_\alpha u) \eta\, dx\,, \quad \forall \eta \in C_0^1(\Omega)\,. \tag{1.2.6}$$

Inserting (1.2.6) into (1.2.4) we get the normal equation

$$\begin{cases} \int_\Omega |\nabla u|^{p-2} \partial_\alpha u \cdot \partial_\alpha \Big(\eta \cdot h_\varepsilon(d(u)) \mathcal{N}(u)\Big) dx \\ \quad = \int_{[u \in \Sigma]} \Theta\, |\nabla u|^{p-2} A_u(\partial_\alpha u, \partial_\alpha u)\, \eta\, dx\,, \\ \eta \in \overset{\circ}{H}{}^{1,p}(\Omega) \cap L^\infty \quad \text{(by approximation)} \end{cases} \tag{1.2.7}$$

(1.2.7) holds for all sufficiently small ε.

Tangential variations:
Let $T: \mathbb{R}^N \to \mathbb{R}^N$ denote a C^1–vectorfield with compact support in a small ball $B_R(z) \subset U$ centered at $z \in \Sigma$ and with the property

$$T(y) \cdot \mathcal{N}(y) = 0\,, \quad \forall y \in B_R(z)\,.$$

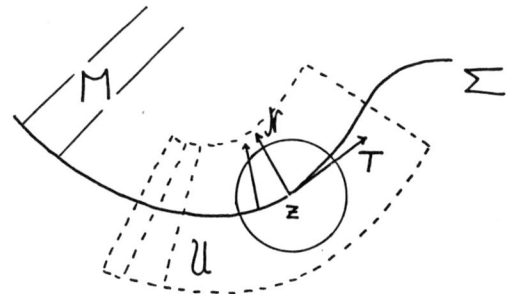

If $\Phi(s,y)$ denotes the flow of T, i.e.

$$\Phi(s,y) := y(s)\,,$$

where $y(s)$ solves

$$\begin{cases} \dot{y}(s) &= T(y(s)) \\ y(0) &= y \end{cases}$$

then

$$v_t := \Phi\Big(t \cdot \eta \cdot h_\varepsilon(d(u))\,,\, u\Big)$$

is admissible for $|t| \ll 1$ and $\eta \in C_0^1(\Omega)$, hence

$$\frac{d}{dt/0} \int_\Omega |\nabla v_t|^p\, dx = 0 \quad \Longrightarrow$$

(observe $\frac{\partial}{\partial t/0} v_t = \eta \cdot h_\varepsilon(d(u)) T(u)$)

$$\begin{cases} \int_\Omega |\nabla u|^{p-2} \partial_\alpha u \cdot \partial_\alpha \Big(\eta \cdot h_\varepsilon(d(u)) T(u)\Big) dx = 0, \\ \eta \in \overset{\circ}{H}{}^{1,p}(\Omega) \cap L^\infty \quad \text{(by approximation)} \end{cases} \quad (1.2.8)$$

Finally we put together (1.2.7) and (1.2.8): We cover $\Sigma \subset \bigcup_{k=1}^{L} B_R(y_k)$ with small balls such that

$$y_k \in \Sigma, \quad B_{3R}(y_k) \subset\subset U$$

and choose a partition of the unity $\{\varphi_k\}_{k=1,\ldots,L}$ such that

$$\operatorname{spt} \varphi_k \subset\subset B_{2R}(y_k),$$

$$\sum_{k=1}^{L} \varphi_k \equiv 1 \text{ on } \bigcup_{k=1}^{L} B_R(y_k) \supset \Sigma.$$

For each $k = 1, \ldots, L$ we select $N-1$ tangential vectorfields $T_{k,1}, T_{k,2}, \ldots, T_{k,N-1}$ such that

$$\operatorname{spt} T_{k,i} \subset\subset B_{3R}(y_k), \quad i = 1, \ldots, N-1, \ k = 1, \ldots, L$$

$$\begin{cases} T_{k,i} \cdot T_{k,j} \equiv \delta_{ij}, \\ T_{k,i} \cdot \mathcal{N} \equiv 0 \end{cases} \quad \text{on } B_{2R}(y_k).$$

For any $\psi \in C_0^1(\Omega, \mathbb{R}^N)$ we deduce

$$\varphi_k(u)\psi = \sum_{i=i}^{N-1} \eta_{k,i}(u) T_{k,i}(u) + \eta_k(u) \mathcal{N}(u)$$

with coefficients

$$\left.\begin{aligned} \eta_{k,i}(u) &= \varphi_k(u) \psi \cdot T_{k,i}(u) \\ \eta_k(u) &= \psi \cdot \mathcal{N}(u) \varphi_k(u) \end{aligned}\right\} \in \overset{\circ}{H}{}^{1,p}(\Omega) \cap L^\infty.$$

Hence (1.2.7) applies with $\eta = \eta_k(u)$ and (1.2.8) is valid with $\eta T(u)$ replaced by $\eta_{k,i}(u) \cdot T_{k,i}(u)$. Adding these results and summing w.r.t. i gives

$$\int_\Omega |\nabla u|^{p-2} \partial_\alpha u \cdot \partial_\alpha \Big(h_\varepsilon(d(u)) \varphi_k(u)\psi\Big) dx$$

$$= \int_{[u \in \Sigma]} \Theta |\nabla u|^{p-2} A_u(\partial_\alpha u, \partial_\alpha u) (\psi \cdot \mathcal{N}(u)) \varphi_k(u) dx.$$

1.2 Linearisation of the Minimum Property, Extension of Maps   15

For small enough ε we can arrange

$$h_\varepsilon(d(y)) = 0 \quad \text{if } y \notin \bigcup_{k=1}^{L} B_R(y_k),$$

hence

$$\sum_{k=1}^{L} \varphi_k(u) = 1 \quad \text{a.e. on } [h_\varepsilon(du) \neq 0],$$

especially

$$\sum_{k=1}^{L} \varphi_k(u) = 1 \quad \text{on } [u \in \Sigma].$$

So taking the sum over $k = 1, \ldots, L$ we deduce

$$\int_\Omega |\nabla u|^{p-2} \partial_\alpha u \cdot \partial_\alpha \Big(h_\varepsilon(d(u)) \cdot \psi \Big) dx$$

$$= \int_{[u \in \Sigma]} \Theta \, |\nabla u|^{p-2} A_u(\partial_\alpha u, \partial_\alpha u) \, \psi \cdot \mathcal{N}(u) \, dx.$$

Finally,

$$\frac{d}{dt/0} \int_\Omega \Big| \nabla \big(u + t \cdot [1 - h_\varepsilon(d(u))] \psi \big) \Big|^p dx = 0,$$

so that

$$\int_\Omega |\nabla u|^{p-2} \partial_\alpha u \cdot \partial_\alpha \psi \, dx = \int_{[u \in \Sigma]} \Theta \, |\nabla u|^{p-2} A_u(\partial_\alpha u, \partial_\alpha u) \psi \cdot \mathcal{N}(u) \, dx$$

is established for $\psi \in C_0^1(\Omega, \mathbb{R}^N)$ or $\in \mathring{H}^{1,p}(\Omega, \mathbb{R}^N) \cap L^\infty$ by approximation.

With some technical modifications which can be found in [30, Theorem 2.1] the following general result can be proved.

Theorem 1.2.1 (Euler equations)

a) Euclidean case: *Suppose that $\Omega \subset \mathbb{R}^n$ is bounded and open, $M \subset \mathbb{R}^N$ a bounded C^2 region with interior unit normal $\mathcal{N}$ and consider bounded, symmetric, elliptic coefficients*

$$a_{\alpha\beta} : \overline{\Omega} \times \mathbb{R}^N \to \mathbb{R}, \quad B^{ij} : \overline{\Omega} \times \mathbb{R}^N \to \mathbb{R}$$

of class C^1 to which we associate the splitting functional

$$\int_\Omega \Big(a_{\alpha\beta}(\cdot, u) \, B^{ij}(\cdot, u) \, \partial_\alpha u^i \partial_\beta u^j \Big)^{p/2} dx = \mathcal{F}(u, \Omega).$$

Assume further that $u \in H^{1,p}(\Omega, \mathbb{R}^N)$ minimizes $\mathcal{F}(\cdot, \Omega)$ subject to the constraint $u(x) \in \overline{M}$. Then there exists an $\mathcal{L}^n$-measurable density $\Theta : \Omega \to [0, 1]$ such that for all $\varphi \in \overset{\circ}{H}{}^{1,p} \cap L^\infty(\Omega, \mathbb{R}^N)$

$$\begin{cases}
\displaystyle\int_\Omega p \cdot a(\cdot, u, \nabla u) \, A^{ij}_{\alpha\beta}(\cdot, u) \, \partial_\alpha u^i \partial_\beta \varphi^j \, dx \\
+ \displaystyle\int_\Omega \frac{p}{2} a(\cdot, u, \nabla u) (\partial_{y^\ell} A^{ij}_{\alpha\beta}(\cdot, u)) \, \partial_\alpha u^i \partial_\beta u^j \, \varphi^\ell \, dx \\
= \displaystyle\int_{[u \in \partial M]} \Theta \frac{\mathcal{N}(u) \cdot \varphi}{\mathcal{N}(u) \cdot \mathcal{M}(\cdot, u)} \, p \, a(\cdot, u, \nabla u) \cdot \Big\{ A^{ij}_{\alpha\beta}(\cdot, u) \, \partial_\alpha u^i \, \partial_\beta (\mathcal{M}^j(\cdot, u)) \\
+ \dfrac{1}{2} \cdot (\partial_{y^\ell} A^{ij}_{\alpha\beta})(\cdot, u) \, \partial_\alpha u^i \partial_\beta u^j \, \mathcal{M}^\ell(\cdot, u) \Big\} \, dx \, ,
\end{cases}$$

moreover $\{\ldots\} \geq 0$ a.e. on $[u \in \partial M]$. Here we have abbreviated

$$A^{ij}_{\alpha\beta}(x, y) := a_{\alpha\beta}(x, y) \cdot B^{ij}(x, y),$$

$$a(x, y, Q) := \left(A^{ij}_{\alpha\beta}(x, y) Q^i_\alpha Q^j_\beta \right)^{\frac{p}{2}-1},$$

$$\mathcal{M}(x, y) := B(x, y)^{-1}(\mathcal{N}(y)).$$

b) *Riemannian case: Let Ω denote a bounded open set in some Riemannian manifold of dimension n and let Y denote an N-dimensional manifold located in some $\mathbb{R}^k$. Assume further that M is a subregion of Y with boundary ∂M of class C^2 such that*

$$M \cup \partial M \subset \mathrm{Int}(Y).$$

If $u \in H^{1,p}(\Omega, \mathbb{R}^k)$ minimizes $\int_\Omega \|du\|^p \, d\,\mathrm{vol}$ under the side condition $u(x) \in M \cup \partial M$ then

$$\int_\Omega p \cdot \|du\|^{p-2} \{du^i \cdot d\psi^i + du^i \cdot du^k \, D^2_{\ell k}\Pi^i(u) \, \psi^\ell\} \, d\,\mathrm{vol}$$

$$= \int_{[u \in \partial M]} p \cdot \Theta \cdot \psi \cdot \nu(u) \, \|du\|^{p-2} \, du^i \cdot d(\nu(u)^i) \, d\,\mathrm{vol}$$

for some density $\Theta : \Omega \to [0, 1]$.

Here ψ is an arbitrary function in $\overset{\circ}{H}{}^{1,p} \cap L^\infty(\Omega, \mathbb{R}^N)$, du^i denote the gradient of u^i w.r.t. the metric on Ω, Π is the smooth nearest point retraction onto Y and ν denotes the interior unit normal of ∂M.

□

1.2 Linearisation of the Minimum Property, Extension of Maps

Remark: In the Euclidean case we considered splitting coefficients $A_{\alpha\beta}^{ij}(x,u) = a_{\alpha\beta}(x,u)B^{ij}(x,u)$. It is an unsolved problem how to obtain an Euler equation similar to the result in part a) of Theorem 1.2.1 if we consider a nonsplitting functional

$$\int_\Omega \left(A_{\alpha\beta}^{ij}(\cdot,u)\, D_\alpha u^i D_\beta u^j\right)^{p/s} dx \,.$$

The problem is unsolved even in case $p = 2$ and $A_{\alpha\beta}^{ij}$ constant. □

We now turn our attention to a more delicate question concerning the construction of suitable comparison functions respecting side conditions of the form $u(x) \in \overline{M}$.

Extension of maps:
For unconstrained local minima of functionals whose integrands satisfy some reasonable growth assumptions *Caccioppoli's inequality* turns out to be a very powerful tool in approaching the regularity of the minimizer since for example Caccioppoli's inequality implies higher integrability of the gradient. The standard device for Caccioppoli's inequality is to insert

$$(1-\eta) \cdot u + \eta \cdot \overline{u}$$

as local comparison function where $\eta : \Omega \to [0,1]$ has compact support and $\overline{u}$ is a mean value. Roughly speaking the test function implying Caccioppoli's inequality is just a *convex combination* of u and $\overline{u}$ which is not admissible when dealing with general obstacle problems. The construction of comparison functions is based on

Theorem 1.2.2 *Suppose that the set M satisfies one of the following hypotheses*

(M1) *M is a bounded open region in Euclidean space $\mathbb{R}^N$ with ∂M of class C^2*

or

(M2) *a smooth bounded open subregion of a k-dimensional manifold $Y \subset \mathbb{R}^N$ with $\overline{M} \subset \operatorname{Int} Y$*

or

(M3) *just a compact submanifold without boundary of $\mathbb{R}^N$.*

Then there exist constants

$$\gamma = \gamma(p) \in (0,1], \quad \varepsilon_0, \delta, q, \tilde{q}, C > 0,$$

depending on dimensions, on p and the geometry of M (and Y in case (M2)) with the following property:

> *If $u \in H^{1,p}(S^{n-1}, \overline{M})$ and $u^* \in \mathbb{R}^N$ are given satisfying*
>
> $$E_p(u, S^{n-1}) \cdot W_p(u, S^{n-1})^\gamma \leq \varepsilon^q \cdot \delta^{\gamma+1}$$
>
> *for some $0 < \varepsilon \leq \varepsilon_0$ then we find an extension $\overline{u} \in H^{1,p}(B^n, \overline{M})$ satisfying the estimates*
>
> $$E_p(\overline{u}, B^n) \leq C \cdot \{\varepsilon \cdot E_p(u, S^{n-1}) + \varepsilon^{-\tilde{q}} W_p(u, S^{n-1})\}$$
> $$W_p(\overline{u}, B^n) \leq C \cdot \varepsilon^{-\tilde{q}} W_p(u, S^{n-1}).$$

Here we have abbreviated

$$E_p(v, A) := \int_A |\nabla v|^p,$$

$$W_p(v, A) := \int_A |v - u^*|^p,$$

$$S^{n-1} := \{x \in \mathbb{R}^n : |x| = 1\},$$

$$B^n := \{x \in \mathbb{R}^n : |x| < 1\}.$$

The quoted version of Theorem 1.2.2 is taken from [31]. Besides all technical details the theorem states:

> *If for some boundary function $\varphi : \partial B^n \to \overline{M}$ the product of energy and mean oscillation is small then φ can be extended to a mapping $\Phi : B^n \to \overline{M}$ whose energy is controlled by a rather large factor times the mean oscillation of φ plus a small contribution of the energy of φ.*

Corollary: If $\varphi \in H^{1,p}(\partial B_r(x), \overline{M})$ satisfies

$$E_p(\varphi, \partial B_r(x)) \cdot W_p(\varphi, \partial B_r(x))^\gamma \leq r^{(1+\gamma) \cdot (n-1) - p} \varepsilon^q \cdot \delta^{\gamma+1}$$

then there exists an extension $\Phi \in H^{1,p}(B_r(x), \overline{M})$ such that

1.2 Linearisation of the Minimum Property, Extension of Maps

$$E_p(\Phi, B_r(x)) \leq C \cdot \{\varepsilon \cdot r \, E_p(\varphi, \partial B_r(x)) + \varepsilon^{-\tilde{q}} r^{1-p} W_p(\varphi, \partial B_r(x))\},$$

$$W_p(\Phi, B_r(x)) \leq C \cdot \varepsilon^{-\tilde{q}} \cdot r \cdot W_p(\varphi, \partial B_r(x)).$$

□

A first version of Theorem 1.2.2 is due to Schoen & Uhlenbeck [71], but their arguments are restricted to the case $p = 2$ and to sets M of type (M3), in [9] we presented the extension theorem for quadratic obstacle problems which is not a corollary to Schoen & Uhlenbeck: in case (M2) for example the $S-U$ extension of $\varphi \in H^{1,2}(S^{n-1}, \overline{M})$ is only in the space $H^{1,2}(B^n, Y)$ but not in the smaller class $H^{1,2}(B^n, \overline{M})$. In the Duzaar & Fuchs paper we overcame this difficulty by first constructing an extension operator

$$H^{1,2}(S^1, \overline{M}) \to H^{1,2}(B^2, \overline{M})$$

from the unit circle to the unit disc and then using the inductive argument of Schoen & Uhlenbeck.

For general $p \neq 2$ we proceed in a similar way: by solving an auxiliary variational problem we first deduce the statement of the theorem for the lower dimensional case $n \leq p$. This step is much more involved than before since we have to prove L^p–estimates for the minimizer of the auxiliary problem.

In order to give some impression of how to proceed we first consider the case $m \leq p$ and a function $u \in H^{1,p}(S^{m-1}, \overline{M})$. Let a denote a fixed point $\in \mathbb{R}^N$. After some technical reductions we can discuss the following case:

$$a \in \Sigma := \partial M \quad \text{and} \quad \underset{S^{m-1}}{\text{osc}} \, u \text{ is small}.$$

(Note that $H^{1,p}(S^{m-1}, \overline{M}) \subset C^0(S^{m-1}, \overline{M})$.

Let $\mathbb{B} := \mathbb{B}_r(a) \subset \Sigma$ denote a geodesic ball in Σ with center a and consider the geodesic cylinder

$$Z := \{y + t\mathcal{N}(y) : y \in \mathbb{B}, \ 0 \leq t \leq h\}$$

of height h which is obtained by moving $\mathbb{B}$ along the direction of the interior unit normal $\mathcal{N}$. Z is constructed to satisfy $\text{Im}(u) \subset Z$. We then look at a solution v of the obstacle problem

$$\int_B |\nabla v|^2 dx \to \min$$

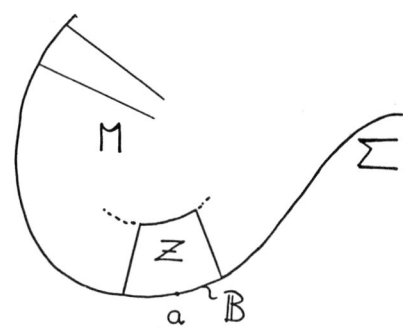

in $H^{1,2}(B, Z)$ for boundary values u, $B := B_1^m(0) \subset \mathbb{R}^m$, and prove:

1. v satisfies (in the sense of distributions) on B

$$-\Delta v = g(\cdot, v, \nabla v)$$

with $|g(\cdot, v, \nabla v)| \leq c(\Sigma) \cdot |\nabla v|^2$. The growth constant $c(\Sigma)$ is independent of v and Z, it only depends on geometric data of Σ. This can be shown by applying a version of Theorem 1.2.1 to the cylinder Z. g is supported on the set $[v \in \partial Z]$. Instead of considering the cylinder Z one could think of replacing Z by $\overline{M} \cap \overline{B}_R(a)$ for a small ball in $\mathbb{R}^N$ but in this case the growth constant will depend on R. For details we refer to [31] and [9].

2. Using the system for v as well as the smallness of $|v - a|$ one can prove

$$\Delta |v - a|^p \geq 0$$

so that

$$\int_B |v - a|^p dx \leq c \cdot \int_{S^{m-1}} |u - a|^p.$$

This is the estimate for the mean oscillation of v.

3. For $r \in [1/2, 1)$ we let

$$v_r(x) := a + \eta_r(|x|) \left[u\left(\frac{x}{|x|}\right) - a \right], \quad x \in B,$$

$$\eta_r(t) := \begin{cases} 0, & t \leq \frac{1}{2}(3r - 1) \\ 1, & t \geq r \\ \text{linear}, & \frac{1}{2}(3r - 1) \leq t \leq r \end{cases}$$

Then $v_r \in H^{1,p}(B, \mathbb{R}^N)$, $v_r = u$ on S^{m-1}, and

$$E_p(v_r, B) \leq c \cdot \{(1-r) E_p(u, S^{m-1}) + (1-r)^{1-p} W_p(u, S^{m-1})\}$$

by direct calculation. Using potential theory (up to the boundary) we then prove

$$\nabla v \in L^p(B), \quad E(v, B) \leq c \cdot E_p(v_r, B)$$

for any r as above. If we choose "$r = 1 - \varepsilon$" the energy estimate of Theorem 1.2.2 will follow.

1.2 Linearisation of the Minimum Property, Extension of Maps

Up to now we have considered boundary maps u with values located in a small cylinder Z. Still assuming $m \leq p$ we now let $u \in H^{1,p}(S^{m-1}, \overline{M})$ denote a general boundary map s.t.

$$E_p(u) \, W_p(u)^{\frac{1}{p-1}} =: \delta^{\frac{p}{p-1}}$$

where δ has to be sufficiently small.

Case 1: $W_p(u) \geq \delta$
implies smallness of $E_p(u)$ and by the way small oscillation of u. Then we choose $\overline{u}$ as the retraction of the p–harmonic extension of u onto $\overline{M}$.

Case 2: $W_p(u) \leq \delta$ but $\text{dist}(a, \Sigma)$ "large".
Then $a \in \text{Int}(M)$ and $\text{Im}(u) \subset \text{Int}(M)$. After appropriate choice of the data also the free p–harmonic map h for boundary values u stays in the interior of M and we may choose $\overline{u} = h$.

Case 3: $W_p(u) \leq \delta$, $\text{dist}(a, \Sigma)$ "small". Then $\text{Im}(u) \subset Z$ for some cylinder Z as above (with base point $a' \in \Sigma$) and we may proceed as before.

Let us remark that for $m \leq p$ Theorem 1.2.2 takes a nicer form:

$\exists \, C, \delta_1, \gamma > 0$ as follows:

$$E_p(u, S^{m-1}) \, W_p(u, S^{m-1})^\gamma \leq \delta_1^{1+\gamma}$$

implies the existence of an extension $\overline{u} \in H^{1,p}(B^m, \overline{M})$ such that

$$E_p(\overline{u}, B^m) \leq \varepsilon \cdot E_p(u, S^{m+1}) + C \cdot \varepsilon^{1-p} \, W_p(u, S^{m-1})$$
$$W_p(\overline{u}, B^m) \leq C \, W_p(u, S^{m-1})$$

holds for all $0 < \varepsilon < 1$.

The inductive step giving Theorem 1.2.2 for all dimensions is very technical and uses similar arguments as in Schoen & Uhlenbeck's paper. (During the inductive procedure one stays in the prescribed set $\overline{M}$ provided the lower dimensional version of the Theorem is proved.)

1.3 Proofs of the Basic Theorems

Here we indicate how to prove some of the results concerning the regularity of constrained minimizers. Partial Hölder continuity for example is based on the following ingredients

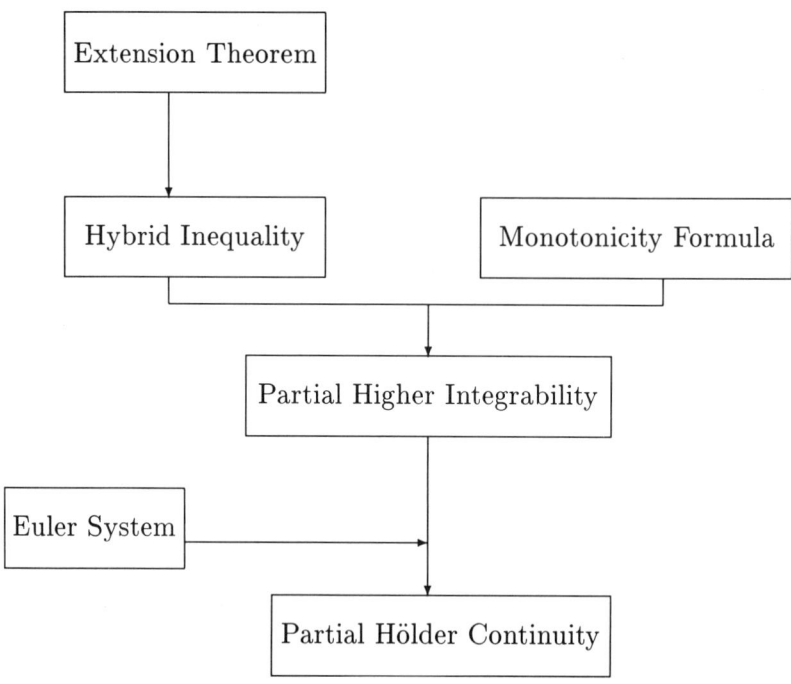

Theorem 1.3.1 (Hybrid Inequality) *Let M denote an arbitrary region as described in Theorem 1.2.2 and suppose that $D \subset \mathbb{R}^n$ is a bounded domain. Let $f_0 : D \times \mathbb{R}^N \times \mathbb{R}^{nN} \to R$ denote a Caratheodory function with the growth properties*

$$k_0|Q|^p \leq f(x, y, Q) \leq k_1|Q|^p$$

for positive constants k_0, k_1. Then if $v \in H^{1,p}(D\overline{M})$ locally minimizes

$$w \mapsto \int_D f(\cdot, w, \nabla w)\, dx$$

in the class $H^{1,p}(D, \overline{M})$ there is a constant C_1 with the following property:

1.3 Proofs of the Basic Theorems

If $\mathcal{E}(v, B_R(x)) := R^{p-n} \int_{B_R(x)} |\nabla v|^p \, dz \leq \frac{1}{C_1} \cdot \lambda^{q/(\gamma+1)}$ for some $0 < \lambda < 1$ then

$$\mathcal{E}(v, B_{R/2}(x)) \leq \lambda \mathcal{E}(v, B_R(x)) + C_1 \cdot \lambda^{-\tilde{q}} \fint_{B_R(x)} |v - (v)_R|^p \, dz .$$

Here $\gamma, q, \tilde{q}$ are the constants appearing in Theorem 1.2.2.

Remarks:

1. Theorem 1.3.1 is a Caccioppoli type inequality but being valid only near points with small energy.

2. If u is a constrained minimizer of $\int_\Omega \|dv\|^p \, d\text{vol}$ defined on some Riemannian manifold Ω then after introducing local coordinates on Ω the representative v of u locally minimizes a functional of the form $\int_D f(\cdot, v, \nabla v) \, dx$.

3. The notion of a "Hybrid Inequality" was first introduced by Hardt & Kinderlehrer & Lin [52] but restricted to quadratic problems for functions into Riemannian manifolds without boundary.

Proof of Theorem 1.3.1: We assume $x = 0$ and denote all absolute constants by $c_1, c_2, \ldots$. Suppose $0 < \lambda < 1$ is given and that

$$\mathcal{E}(v, B_R) \leq \frac{1}{C_1} \cdot \lambda^{q/(\gamma+1)}$$

holds with C_1 specified later. According to Fubini's Theorem (compare [68], Theorem 3.6.1(c)) there is a radius $r \in [\frac{R}{2}, R]$ such that

$$E_p(v, \partial B_r) \leq \frac{8}{R} E_p(v, B_R),$$

$$W_p(v, B_r) \leq \frac{8}{R} W_p(v, B_R)$$

$W_p(v, \cdot)$ being caculated w.r.t. $(v)_R = \fint_{B_R} v \, dx$. This gives on account of Poincaré's inequality

$$r^{p-(n-1)(\gamma+1)} E_p(v, \partial B_r) \cdot W_p(v, \partial B_r)^\gamma$$
$$\leq c_1 R^{-1-\gamma} \cdot R^{p-(n-1)(\gamma+1)} E_p(v, B_R) W_p(v, B_R)^\gamma$$
$$\leq c_2 \cdot \mathcal{E}(v, B_R)^{1+\gamma} \leq c_2 \cdot C_1^{-1-\gamma} \cdot \lambda^q .$$

Let $\varepsilon := \lambda \cdot \mu$ with μ being determined later. Then the above inequality can be rewritten as

$$r^{p-(n-1)(\gamma+1)} E_p(v, \partial B_r) \cdot W_p(v, \partial B_r)^\gamma \leq (c_2 \cdot C_1^{-1-\gamma} \mu^{-q}) \varepsilon^q$$

and if we assume (δ taken from Theorem 1.2.2)

$$c_2 \cdot C_1^{-1-\gamma} \mu^{-q} \leq \delta^{\gamma+1} \iff C_1 \geq (c_2 \mu^{-q})^{1/(1+\gamma)} \cdot \delta^{-1} \qquad (1.3.1)$$

Theorem 1.2.2 implies the existence of $\bar{v} \in H^{1,p}(B_r, \overline{M})$ with boundary values v and the energy estimate

$$E_p(\bar{v}, B_r) \leq C\{\varepsilon \cdot R \cdot E_p(v, \partial B_r) + \varepsilon^{-\tilde{q}} R^{1-p} W_p(v, \partial B_r)\}.$$

Since v is a local minimizer of $\int f(\cdot, v, \nabla v)\, dx$ we deduce by using $\bar{v}$ as comparison function

$$\int_{B_r} f(\cdot, v, \nabla v)\, dx \leq \int_{B_r} f(\cdot, \bar{v}, \nabla \bar{v})\, dx,$$

hence

$$\begin{aligned}
\mathcal{E}(v, B_{R/2}) &\leq c_3 \cdot k_0^{-1} R^{p-n} \int_{B_r} f(\cdot, v, \nabla v)\, dx \\
&\leq c_3 k_0^{-1} R^{p-n} \int_{B_r} f(\cdot, \bar{v}, \nabla \bar{v})\, dx \leq c_4 \frac{k_1}{k_0} R^{p-n} E_p(\bar{v}, B_r) \\
&\leq c_5 \frac{k_1}{k_0} R^{p-n} \{\varepsilon \cdot E_p(v, B_R) + \varepsilon^{-\tilde{q}} R^{-p} W_p(v, B_R)\} \\
&\leq c_5 \frac{k_1}{k_0} \varepsilon \cdot \mathcal{E}(v, B_R) + c_5 \frac{k_1}{k_0} \varepsilon^{-\tilde{q}} \fint_{B_R} |v - (v)_R|^p\, dx.
\end{aligned}$$

Define $\mu := c_5^{-1} \frac{k_0}{k_1}$ (by enlarging c_5 we may assume $\mu \leq \varepsilon_0$, ε_0 taken from Theorem 1.2.2, especially $\varepsilon \leq \varepsilon_0$). Then

$$\mathcal{E}(v, B_{R/2}) \leq \lambda \cdot \mathcal{E}(v, B_R) + \left[c_5 \frac{k_1}{k_0}\right]^{\tilde{q}+1} \lambda^{-\tilde{q}} \cdot \fint_{B_R} |v - (v)_R|^p\, dx$$

and the statement of Theorem 1.3.1 follows if (recall (1.3.1)) we finally choose

$$C_1 := \max\left\{\delta^{-1} \cdot (c_2 \cdot \mu^{-q})^{\frac{1}{1+\gamma}} \cdot \delta^{-1},\ \left[c_5 \cdot \frac{k_1}{k_0}\right]^{\tilde{q}+1}\right\}.$$

□

1.3 Proofs of the Basic Theorems

If the scaled p-energy $R^{p-n} \int_{B_R(x)} |\nabla u|^p \, dz$ is controlled by some power λ^β (with fixed exponent $0 < \beta < 1$) then the scaled energy on the ball of half the radius is dominated by the mean oscillation plus λ times the original energy. (One should think of λ as being very small.)

So the next question which arises concerns the behaviour of the scaled p-energy as a function of the radius. We first recall a

Theorem (of Giusti): Let Ω denote an open set in $\mathbb{R}^n$ and consider a function $w \in H^{1,p}(\Omega)$. Then

$$\mathcal{H}^{n-p}\left(\left\{x \in \Omega : \limsup_{R \downarrow 0} R^{p-n} \int_{B_R(x)} |\nabla w|^p \, dz > 0\right\}\right) = 0.$$

Note that w is arbitrary, i.e. not necessarily minimizing a variational integral. Minimality is basic for the next result.

Theorem 1.3.2 (Monotonicity formula) *Suppose that $\Omega \subset \mathbb{R}^n$ is an open set and that $u \in H^{1,p}(\Omega, \overline{M})$ locally minimizes $\int_\Omega |\nabla u|^p \, dx$ in the constrained class. Then we have ($p \leq n$)*

$$r^{p-n} \int_{B_r(x)} |\nabla u|^p \, dz - \rho^{p-n} \int_{B_\rho(x)} |\nabla u|^p \, dz$$

$$= p \cdot \int_{B_r(x) - B_\rho(x)} |x - z|^{p-n} |\nabla u|^{p-2} |\partial_r u|^2 \, dz$$

for all balls $B_\rho(x) \subset B_r(x) \subset\subset \Omega$. Here $\partial_r u$ denotes the radial derivative of u. Especially, $t \mapsto t^{p-n} \int_{B_t(x)} |\nabla u|^p \, dz$ is increasing.

Remarks: Similar formulas hold for minimizers of

$$\int_\Omega \left(a_{\alpha\beta}(x) B^{ij}(x,u) \partial_\alpha u^i \partial_\beta u^j\right)^{p/2} dx$$

with coefficients $a_{\alpha\beta}$ independent of u, we omit the details and refer to [43] or [44, Lemma 3.2, p. 262] where one has to replace the exponent. For the readers' convenience we give

Some ideas of the proof: Suppose $x = 0$ and consider the comparison function ($0 < t < 1$)

$$u_t(x) = \begin{cases} u(x), & |x| \geq t \\ u\left(t \frac{x}{|x|}\right), & |x| \leq t \end{cases}$$

which is a radial deformation in the domain of definition without changing the image space. From

$$\int_{B_t} |\nabla u|^p \, dz \leq \int_{B_t} |\nabla u_t|^p \, dz$$

one easily gets (by calculation)

$$\int_{B_t} |\nabla u|^p \, dz \leq \frac{t}{n-p} \left\{ \int_{\partial B_t} |\nabla u|^p \, d\mathcal{H}^{n-1} - \int_{\partial B_t} |\partial_r u|^2 \cdot |\nabla u|^{p-2} \, d\mathcal{H}^{n-1} \right\}.$$

Let $\Phi(t) := t^{p-n} \int_{B_t} |\nabla u|^p \, dx$. Then

$$\begin{aligned}
\Phi'(t) &= (p-n) t^{p-n-1} \int_{B_t} |\nabla u|^p \, dx + t^{p-n} \int_{\partial B_t} |\nabla u|^p \, d\mathcal{H}^{n-1} \\
&\geq t^{p-n} \int_{\partial B_t} |\partial_r u|^2 \, |\nabla u|^{p-2} \, d\mathcal{H}^{n-1}
\end{aligned}$$

and after integration

$$\begin{aligned}
\Phi(r) - \Phi(\rho) &\geq \int_\rho^r t^{p-n} \int_{\partial B_t} |\partial_r u|^2 \, |\nabla u|^{p-2} \, d\mathcal{H}^{n-1} \\
&= \int_{B_r - B_\rho} |x|^{p-n} \, |\partial_r u|^2 \cdot |\nabla u|^{p-2} \, dx.
\end{aligned}$$

□

With a more careful choice of the comparison function one can actually prove

$$\begin{aligned}
\Phi(r) - \Phi(\rho) &= p \cdot \int_{B_r - B_\rho} \cdots \\
&\geq \int_{B_r - B_\rho} \cdots .
\end{aligned}$$

in place of

The details can be found in [53]. □

We now combine the monotonicity formula and the hybrid inequality in order to obtain higher integrability of the gradient of a local minimizer near a point with sufficiently small energy.

Theorem 1.3.3 (Partial higher integrability) *There exists $\varepsilon_1 > 0$, $t > p$ and $c > 0$ as follows: If $u \in H^{1,p}(\Omega, \overline{M})$ minimizes $\int_\Omega |\nabla u|^p \, dx$ and if*

1.3 Proofs of the Basic Theorems

$$\Phi\bigl(u, B_{R_0}(x_0)\bigr) := R_0^{p-n} \int_{B_{R_0}(x_0)} |\nabla u|^p \, dx < \varepsilon_1^p$$

for some ball $B_{R_0}(x_0) \subset \Omega$ *then* $\nabla u \in L^t_{\mathrm{loc}}\bigl(B_{R_0}(x_0)\bigr)$ *and*

$$\left(\fint_{B_{r/2}(x)} |\nabla u|^t \, dz\right)^{1/t} \le c \cdot \left(\fint_{B_r(x)} |\nabla u|^p \, dz\right)^{1/p}$$

for all $x \in B_{R_0/2}(x_0)$, $r \in (0, \frac{R_0}{2})$.

Remark. The same result is true for more general functionals, e.g. splitting functionals with $a_{\alpha\beta} = a_{\alpha\beta}(x)$ (so that the monotonicity formula holds).

Proof: Suppose

$$\Phi(u, B_{R_0}(x_0)) < \varepsilon^p \tag{1.3.2}$$

for some small $\varepsilon > 0$ being determined later. We apply the monotonicity formula to the ball $B_r(x)$, $x \in B_{R_0/2}(x_0)$, $r \in (0, \frac{R_0}{2})$ and get

$$\begin{aligned}
\Phi(u, B_r(x)) &\le \Phi\bigl(u, B_{R_0/2}(x)\bigr) = 2^{n-p} R_0^{p-n} \int_{B_{R_0/2}(x)} |\nabla u|^p \, dz \\
&\le 2^{n-p} R_0^{p-n} \int_{B_{R_0}(x_0)} |\nabla u|^p \, dx \le 2^{n-p} \varepsilon^p
\end{aligned}$$

by assumption (1.3.2). With the notions from Theorem 1.3.1 let

$$\lambda := \left(C_1 \, 2^{n-p} \varepsilon^p\right)^{\frac{\gamma+1}{q}}.$$

Then

$$\Phi(u, B_r(x)) \le \lambda^{\frac{q}{\gamma+1}} \frac{1}{C_1}$$

and the hybrid inequality implies

$$\Phi(u, B_{r/2}(x)) \le \lambda \cdot \Phi(u, B_r(x)) + C_1 \lambda^{-\tilde{q}} \fint_{B_r(x)} |u - (u)_r|^p \, dz$$

$$\Longrightarrow \fint_{B_{r/2}(x)} |\nabla u|^p \, dx \le c(n,p) \lambda \cdot \fint_{B_r(x)} |\nabla u|^p \, dx$$

$$+ C_1 \cdot c(n,p) \lambda^{-\tilde{q}} r^{-p} \cdot \fint_{B_r(x)} |u - (u)_r|^p \, dz.$$

We fix $\lambda := \frac{1}{2} c(n,p)^{-1}$ or equivalently

$$\varepsilon_1 := \left(\frac{1}{2}c(n,p)^{-1}\right)^{\frac{q}{p\cdot\gamma+p}} \cdot C_1^{-\frac{1}{p}} \left(2^{n-p}\right)^{-\frac{1}{p}}$$

and arrive at

$$\fint_{B_{r/2}(x)} |\nabla u|^p \, dx \leq \frac{1}{2} \fint_{B_r(x)} |\nabla u|^p \, dx + \kappa \cdot r^{-p} \fint_{B_r(x)} |u - (u)_r|^p \, dx \quad (1.3.3)$$

for all $x \in B_{R_0/2}(x_0)$, $r \in (0, \frac{R_0}{2})$ provided (1.3.2) holds with $\varepsilon = \varepsilon_1$. κ is a positive absolute constant. Applying [44, Prop. V1.1] to inequality (1.3.3) the statements of Theorem 1.3.3 are immediate. □

After these preparations we now give a

Proof of Theorem A: In order to keep things simple we again assume that $u \in H^{1,p}(\Omega, \overline{M})$ is a local minimizer of $\int_\Omega |\nabla u|^p \, dx$ in the constrained space. Let $B_{R_0}(x_0)$ denote a ball in Ω such that

$$R_0^{p-n} \int_{B_{R_0}(x_0)} |\nabla u|^p \, dx < \varepsilon^p \quad (1.3.4)$$

for some $\varepsilon \in (0, \varepsilon_1]$, ε_1 defined in Theorem 1.3.3. From Theorem 1.2.1 we deduce

$$\int_\Omega |\nabla u|^{p-2} \nabla u \cdot \nabla \varphi \, dx = \int_\Omega f(\cdot, u, \nabla u) \cdot \varphi \, dx \quad (1.3.5)$$

for all $\varphi \in H^{1,p} \cap L^\infty(\Omega, \mathbb{R}^N)$ with compact support in Ω. The explicit form of the right-hand-side f is of minor interest, we only use the growth property

$$|f(\cdot, u, \nabla u)| \leq c_1 |\nabla u|^p \quad (1.3.6)$$

for some absolute constant c_1. Let $B_r(\tilde{x})$ denote a ball with center $\tilde{x} \in B_{R_0/2}(x_0)$ and radius $r \in (0, \frac{R_0}{2})$ and consider the solution $v \in H^{1,p}\left(B_{r/2}(\tilde{x}), \mathbb{R}^N\right)$ of

$$\int_{B_{r/2}(\tilde{x})} |\nabla v|^p \, dx \to \min$$

in $u + \overset{\circ}{H}^{1,p}\left(B_{r/2}(\tilde{x}), \mathbb{R}^N\right)$. Then $\varphi := u - v$ is admissible in (1.3.5) and by observing

$$(A|A|^{p-2} - B|B|^{p-2}) \cdot (A - B) \geq c_2 |A - B|^p, \quad A, B \in \mathbb{R}^{nN}$$

we deduce from (1.3.5), (1.3.6)

1.3 Proofs of the Basic Theorems

$$\fint_{B_{r/2}(\tilde{x})} |\nabla u - \nabla v|^p \, dx \leq c_3 \fint_{B_{r/2}(\tilde{x})} |u - v| \cdot |\nabla u|^p \, dx. \qquad (1.3.7)$$

The integral on the right-hand-side of (1.3.7) is handled with the reverse Hölder-inequality proven in Theorem 1.3.3:

$$\fint_{B_{r/2}(\tilde{x})} |\nabla u|^p \, |u - v| \, dx$$

$$\leq \left(\fint_{B_{r/2}(\tilde{x})} |\nabla u|^t \, dx \right)^{p/t} \cdot \left(\fint_{B_{r/2}(\tilde{x})} |u - v|^{\frac{t}{t-p}} \, dx \right)^{1-p/t}$$

$$\leq c_4 \cdot \fint_{B_r(\tilde{x})} |\nabla u|^p \, dx \cdot \left(\fint_{B_{r/2}(\tilde{x})} |u - v|^{\frac{t}{t-p}} \, dx \right)^{1-p/t}$$

$$\leq c_5 \cdot \fint_{B_r(\tilde{x})} |\nabla u|^p \, dx \cdot \left(\fint_{B_{r/2}(\tilde{x})} |u - v|^p \, dx \right)^{1-p/t}$$

where we have used the boundedness of u and v (observe here $\sup_{B_{r/2}(\tilde{x})} |v| \leq \sup_{\partial B_{r/2}(\tilde{x})} |u|$, $\mathrm{Im}\,(u) \subset \overline{M} = $ compact). Finally we apply Poincaré's inequality to end up with

$$\fint_{B_{r/2}(\tilde{x})} |\nabla u|^p \cdot |u - v| \, dx$$

$$\leq c_6 \cdot \fint_{B_r(\tilde{x})} |\nabla u|^p \cdot \left(r^p \fint_{B_{r/2}(\tilde{x})} |\nabla u - \nabla v|^p \, dx \right)^{1-p/t} \qquad (1.3.8)$$

$$\leq c_7 \fint_{B_r(\tilde{x})} |\nabla u|^p \, dx \cdot \left\{ r^p \fint_{B_r(\tilde{x})} |\nabla u|^p \, dx \right\}^{1-p/t}$$

on account of the minimality of v. For $0 < \tau < \frac{1}{4}$ we have

$$\Phi(u, B_{\tau r}(\tilde{x})) := (\tau r)^{p-n} \int_{B_{\tau r}(\tilde{x})} |\nabla u|^p \, dx$$

$$\leq c_8 \left\{ (\tau r)^{p-n} \int_{B_{\tau r}(\tilde{x})} |\nabla v|^p \, dx + (\tau r)^{p-n} \int_{B_{r/2}(\tilde{x})} |\nabla u - \nabla v|^p \, dx \right\}$$

$$\underset{(1.3.7),(1.3.8)}{\leq} c_9 \left\{ \Phi(v, B_{\tau r}(\tilde{x})) + (\tau r)^{p-n} \left(\int_{B_r(\tilde{x})} |\nabla u|^p \, dx \right) \right.$$

$$\left. \cdot \left[r^p \cdot \fint_{B_r(\tilde{x})} |\nabla u|^p \, dx \right]^{1-p/t} \right\}.$$

A theorem of Uhlenbeck [80] states

$$\sup_{B_{\tau r}(\tilde{x})} |\nabla v|^p \leq c_{10} \fint_{B_{r/2}(\tilde{x})} |\nabla v|^p \, dx \,,$$

so that

$$\begin{aligned}\Phi(v, B_{\tau r}(\tilde{x})) &\leq c_{11} \tau^p \, \Phi(v, B_{r/2}(\tilde{x})) \\ &\leq c_{12} \tau^p \, \Phi(u, B_{r/2}(\tilde{x})) \\ &\leq c_{13} \cdot \tau^p \cdot \Phi(u, B_r(\tilde{x}))\end{aligned}$$

by quoting the minimality of v one more time. Therefore

$$\Phi(u, B_{\tau r}(\tilde{x})) \leq c_{14} \tau^p \cdot \left\{ 1 + \tau^{-n} \Phi(u, B_r(\tilde{x}))^{1-p/t} \right\} \cdot \Phi(u, B_r(\tilde{x})) \,.$$

Next observe (Theorem 1.3.2)

$$\Phi(u, B_r(\tilde{x})) \leq \Phi(u, B_{R_0/2}(\tilde{x})) \leq c_{15} \Phi(u, B_{R_0}(x_0))$$

so that

$$\tau^{-n} \Phi(u, B_r(\tilde{x}))^{1-p/t} \leq c_{16} \tau^{-n} \, \Phi(u, B_{R_0}(x_0))^{1-p/t} \,.$$

If we thus let $\tau := \left(\frac{1}{4} \cdot c_{14}^{-1} \right)^{1/p}$ and require in addition to (1.3.4)

$$\Phi(u, B_{R_0}(x_0)) \leq \left(c_{16}^{-1} \tau^n \right)^{\frac{t}{t-p}} \tag{1.3.9}$$

then $\tau^{-n} \Phi(u, B_r(\tilde{x}))^{1-p/t} \leq 1$ and we have shown

$$\Phi(u, B_{\tau r}(\tilde{x})) \leq \frac{1}{2} \cdot \Phi(u, B_r(\tilde{x}))$$

for all $\tilde{x} \in B_{R_0/2}(x_0)$, $r \in \left(0, \frac{R_0}{2} \right)$. We apply this result inductively to $r_k := \frac{1}{2} \tau^k R_0$ and get

$$\Phi(u, B_{r_k}(\tilde{x})) \leq 2^{-k} \Phi(u, B_{R_0/2}(\tilde{x})) \underset{(1.3.9)}{\leq} 2^{-k} \tau^{n \cdot t/(t-p)}$$

Now, if $0 < r \leq R_0/2$ we calculate k such that

$$r_k \leq r < r_{k-1}$$

and deduce from the above inequality

$$\Phi(u, B_r(\tilde{x})) \leq \tau^{\frac{nt}{t-p}} \cdot 2 \cdot \left\{ \frac{2r}{R_0} \right\}^{-\log 2 / \log \tau} \,.$$

1.3 Proofs of the Basic Theorems

Due to the arbitrariness of $\tilde{x}$ and r Morrey's Dirichlet Growth-Theorem applies:

$$\begin{cases} u \in C^{0,\alpha}(B_{R_0/2}(x_0)), \quad \alpha = -\frac{1}{p} \cdot \frac{\log 2}{\log \tau} \\ \text{and} \\ |u(x) - u(y)| \leq c(R_0) \cdot |x-y|^\alpha. \end{cases} \quad (1.3.10)$$

On the other hand an easy calculation gives using (1.3.5)

$$x_0 \in \text{Reg}(u) \implies \limsup_{r \downarrow 0} \Phi(u, B_r(x_0)) = 0.$$

Thus we have shown:

$$\text{Reg}(u) = \left\{ x \in \Omega : \liminf_{r \downarrow 0} \Phi(u, B_r(x)) = 0 \right\}. \quad (\Rightarrow \mathcal{H}^{n-p}(\text{Sing } u) = 0).$$

It is another easy exercise to prove $u \in C^{0,\beta}(\text{Reg } u)$ for any $0 < \beta < 1$, compare [30, p. 145–146].

Remarks:

1. It should be noted explicitly that there exists an ε_2 depending on absolute data such that

$$R_0^{p-n} \int_{B_{R_0}(x_0)} |\nabla u|^p \, dx < \varepsilon_2^p$$

   for some ball $B_{R_0}(x_0)$ implies $u \in C^{0,\beta}(B_{R_0/2}(x_0))$ for all $0 < \beta < 1$. Moreover, we have the apriori Hölder bound

$$|u(x) - u(y)| \leq c(R_0, \beta, \ldots) \cdot |x-y|^\beta$$

   on $B_{R_0/2}(x_0)$.

2. With technical difficulties the proof extends to more general functionals (of splitting type) including also the Riemannian case.

□

We give a short outline of how to prove the

Optimal estimate for the interior singular set: We make use of the so-called dimension reduction technique which is a powerful tool in Geometric Measure Theory (already used by Federer in the sixties) and which in recent years has been successfully applied to variational problems for vector valued functions. The basic idea can be summarized as follows:

Suppose that u is a minimizer with isolated singularity at 0. Scaling $u_\lambda(x) := u(\lambda x)$, $|x| < 1$, leads to a sequence of local minimizers u_λ which converge weakly to some limit function u_0. One must show:

- u_0 is as regular as a minimizing map

- u_0 is radially independent

Then the analysis of the singular set of the tangent map u_0 gives additional information on $\mathrm{Sing}(u)$.

In order to carry out the first step of the above program one has to replace the partial regularity criterion "$\Phi(u, B_r(x)) < \varepsilon$" by a condition being stable under weak convergence.

Lemma 1.3.1 *Suppose $B > 0$ and a point $u^* \in \mathbb{R}^N$ are given. There are constants $\varepsilon_3 = \varepsilon_3(n, N, p, B, M)$, $c = c(n, N, M, p)$, $\alpha = \alpha(n, N, M, p)$ as follows: If $u \in H^{1,p}(B_1, \overline{M})$ locally minimizes $\int_{B_1} |\nabla u|^p \, dx$ and if*

$$\Phi(u, B_1) \leq B, \; W_p(u, B_1) := \int_{B_1} |u - u^*|^p \, dx < \varepsilon_3^p$$

then $u \in C^{0,\alpha}(B_{1/2}, \mathbb{R}^N)$ and $|u(x) - u(y)| \leq c \cdot |x - y|^\alpha$, $x, y \in B_{1/2}$.

Remarks:

1. The Lemma roughly states that we have local regularity at points with small mean oscillation. The actual size of this ε-bound for W_p depends on the associated scaled energy which is not required to be small. For problems with a Caccioppoli inequality smallness of W_p immediately implies smallness of the scaled energy. Here we make use of the extension theorem which is applicable since the product of scaled energy and mean oscillation is small.

2. Note that c and α are independent of B.

3. Clearly Lemma 1.3.1 extends to more complicated functionals but we try to avoid notational difficulties.

Proof of Lemma 1.3.1: The first remark after the proof of Theorem A shows that $u \in C^{0,\alpha}(B_{1/2})$, $|u(x) - u(y)| \leq c|x - y|^\alpha$ on $B_{1/2}$, with c and α as stated provided

$$\Phi(u, B_{3/4}) < \varepsilon_2^p \tag{1.3.11}$$

for a certain $\varepsilon_2 = \varepsilon_2(n, N, M, p)$. In what follows we denote by $c_1, c_2 \ldots$ constants depending only on n, N, M, p. Select a radius $r \in [\frac{3}{4}, 1]$ such that

$$\begin{aligned} E_p(u, \partial B_r) &\leq 8 E_p(u, B_1), \\ W_p(u, \partial B_r) &\leq 8 W_p(u, B_1) \end{aligned}$$

1.3 Proofs of the Basic Theorems

and apply Theorem 1.2.2 to find an extension $\bar{u} \in H^{1,p}(B_r, \overline{M})$ such that

$$E_p(\bar{u}, B_r) \leq C \cdot \{\varepsilon \cdot E_p(u, \partial B_r) + \varepsilon^{-\tilde{q}} W_p(u, \partial B_r)\}.$$

This estimate is valid provided

$$E_p(u, \partial B_r) \cdot W_p(u, \partial B_r)^\gamma \leq \varepsilon^q \cdot \delta^{\gamma+1} \tag{1.3.12}$$

where up to now ε is arbitrary and $C, q, \tilde{q}, \delta, \gamma$ are defined in Theorem 1.2.2. From the choice of r we deduce

$$E_p(u, \partial B_r) \cdot W_p(u, \partial B_r)^\gamma \leq c_1 \cdot B \cdot \varepsilon_3^{p \cdot \gamma}$$

with ε_3 being defined later. Therefore (1.3.12) follows from

$$c_1 \cdot B \cdot \varepsilon_3^{p \cdot \gamma} \leq \varepsilon^q \cdot \delta^{\gamma+1}. \tag{1.3.13}$$

The minimality of u implies

$$\begin{aligned} E_p(u, B_r) \leq E_p(\bar{u}, B_r) &\leq C \cdot (8 \cdot B \cdot \varepsilon + \varepsilon^{-\tilde{q}} \cdot \varepsilon_3^q) \\ &= c_2 \cdot (B \cdot \varepsilon + \varepsilon^{-\tilde{q}} \cdot \varepsilon_3^q). \end{aligned} \tag{1.3.14}$$

We choose

$$\varepsilon = \varepsilon(n, N, M, B, p)$$

to satisfy (ε_2 from (1.3.11))

$$c_2 \cdot B \cdot \varepsilon \leq \frac{1}{2} \cdot \varepsilon_2^p \tag{1.3.15}$$

and then fix ε_3 according to

$$\varepsilon_3 \leq \min\left(\left[c_1^{-1} \cdot B^{-1} \varepsilon^q \cdot \delta^{\gamma+1}\right]^{\frac{1}{p\gamma}}, \left[c_2^{-1} \varepsilon^{\tilde{q}} \frac{1}{2} \varepsilon_2^p\right]^{1/q}\right). \tag{1.3.16}$$

Then (1.3.14), (1.3.15), (1.3.16) imply $E_p(u, B_r) \leq \varepsilon_2^p$ which is (1.3.11) up to another constant. $\square$

The next lemma is the desired compactness property of sequences of minimizing functions which will also be of importance for further applications.

Lemma 1.3.2 Let $\{u_i\} \subset H^{1,p}(B_1, \overline{M})$ denote a sequence of local minimizers of $\int_{B_1} |\nabla w|^p \, dx$ such that $u_i \rightharpoonup u$ weakly in $H^{1,p}(B_1, \mathbb{R}^N)$ for some function u in this space. Then there exists a relatively closed subset Σ of B_1 such that $\mathcal{H}^{n-p}(\Sigma) = 0$ and $u \in C^{0,\alpha}(B_1 - \Sigma)$ for some $\alpha \in (0,1)$. Moreover, we have strong convergence $u_i \to u$ in $H^{1,p}_{loc}(B_1, \mathbb{R}^N)$ and uniform convergence on compact subsets of $B_1 - \Sigma$.

Remark: It is easy to see that the limit function u locally minimizes $\int_{B_1} |\nabla w|^p \, dx$ on compact subsets of the regular set $B_1 - \Sigma$. Luckhaus [67] recently showed that u is a local minimizer w.r.t. the whole ball but we will not make use of this fact which is rather hard to prove. *Proof:* We have $B := \sup_i \|\nabla u_i\|^p_{L^p(B_1)} < \infty$ and after passing to subsequences we can arrange

$$u_i \to u \text{ strongly in } L^p(B_1, \mathbb{R}^N), \quad u_i \to u \text{ a.e. on } B_1.$$

Define $\Sigma := B_1 - \text{Reg}(u)$ (we do not know $\text{Reg}(u) \neq \emptyset$) and choose $x_0 \in B_{1/2}$ with the property

$$\fint_{B_r(x_0)} |u - (u)_r|^p \, dx < \frac{1}{2} \cdot \varepsilon_3^p \tag{1.3.17}$$

for a ball $B_r(x_0)$ with $\varepsilon_3 = \varepsilon_3(B)$ defined in Lemma 1.3.1. The strong L^p-convergence turns (1.3.17) into the estimate

$$\fint_{B_r(x_0)} |u_i - (u_i)_r|^p \, dx < \varepsilon_3^p \tag{1.3.18}$$

for $i \gg 1$. In terms of the scaled functions

$$U_i(z) := u_i(x_0 + r \cdot z), \quad z \in B_1,$$

(1.3.18) reads

$$\fint_{B_1} |U_i - (U_i)_1|^p \, dz < \varepsilon_3^p, \tag{1.3.18*}$$

moreover we have (assuming $r < \frac{1}{2}$)

$$\begin{aligned}\Phi(U_i, B_1) &= \Phi(u_i, B_r(x_0)) \leq \Phi(u_i, B_{1/2}(x_0)) \\ &\leq 2^{n-p} \int_{B_1} |\nabla u_i|^p \, dx \leq 2^{n-p} \cdot B.\end{aligned} \tag{1.3.19}$$

Obviously U_i is a local minimizer of $\int_{B_1} |\nabla w|^p \, dz$ on B_1 and (1.3.18*), (1.3.19) are just the assumptions of Lemma 1.3.1 (to be precise one has to calculate ε_3 w.r.t. to $2^{n-p}B$ instead of B). Therefore we find $\alpha \in (0,1)$ and $\kappa > 0$ independent of i such that $|U_i(x) - U_i(y)| \leq \kappa |x-y|^\alpha$ on $B_{1/2}$ for all $i \gg 1$ or equivalently

$$|u_i(x) - u_i(y)| \leq \kappa \cdot |x - y|^\alpha, \quad x, y \in B_{r/2}(x_0), \, i \gg 1. \tag{1.3.20}$$

From Arcela's Theorem, (1.3.20) and $u_i \to u$ a.e. we deduce $u \in C^{0,\alpha}(B_{r/2}(x_0))$ and uniform convergence $u_i \to u$ on this ball. In conclusion:

1.3 Proofs of the Basic Theorems

$$\left\{x \in \Omega : \liminf_{r \downarrow 0} \fint_{B_r(x)} |u - (u)_r|^p \, dx = 0\right\} \subset \text{Reg}(u)$$

so that $\mathcal{H}^{n-p}(\Sigma) = 0$ by Giusti's Theorem. Moreover, $u_i \to u$ uniformly on compact subsets of $B_1 - \Sigma$.

Next we prove $\nabla u_i \to \nabla u$ in $L^p(B_{1/2})$ (and a trivial modification of the following argument will give $L^p_{\text{loc}}(B_1)$-convergence). Let $\varepsilon > 0$ be given. $\mathcal{H}^{n-p}(\Sigma) = 0$ implies the existence of a covering $\Sigma \cap B_{1/2} \subset \bigcup_{i=1}^{\infty} B_i$, $B_i := B_{r_i}(x_i)$, such that

$$\sum_{i=1}^{\infty} r_i^{n-p} < \varepsilon.$$

We let $U := \bigcup_{i=1}^{\infty} B_i$. Using the monotonicity formula we deduce the following energy bound:

$$\int_U |\nabla u_j|^p \, dx \leq \sum_{i=1}^{\infty} \int_{B_{r_i}(x_i)} |\nabla u_j|^p \, dx$$

$$= \sum_{i=1}^{\infty} r_i^{n-p} \left(r_i^{p-n} \int_{B_{r_i}(x_i)} |\nabla u_j|^p \, dx\right)$$

$$\leq \sum_{i=1}^{\infty} r_i^{n-p} \left(\frac{1}{2}\right)^{p-n} \int_{B_{1/2}(x_i)} |\nabla u_j|^p \, dx$$

$$\leq 2^{n-p} \cdot B \sum_{i=1}^{\infty} r_i^{n-p}$$

$$\leq 2^{n-p} \cdot B \cdot \varepsilon,$$

and since ε was arbitrary we have $\int_U |\nabla u_j|^p \, dx \xrightarrow[j \to \infty]{} 0$.

Next recall

$$\int_{B_1} |\nabla u_j|^{p-2} \nabla u_j \cdot \nabla \varphi \, dx = \int_{B_1} f_j \cdot \varphi \, dx \qquad (1.3.21)$$

$$\forall \varphi \in H^{1,p} \cap L^\infty(B_1, \mathbb{R}^N), \text{ spt } \varphi \text{ compact}$$

with $|f_j| \leq a \cdot |\nabla u_j|^p$ for some $a > 0$. We write down (1.3.21) for arbitrary j, k and insert $\varphi := \eta^p \cdot (u_j - u_k)$ with $\eta \in C_0^1(B_1, [0, 1])$, spt $\eta \cap \Sigma = \emptyset$, $\eta = 1$ on $\overline{B}_{1/2} - U$. Subtraction of the results gives

$$\int \eta^p |\nabla u_j - \nabla u_k|^p \, dx \leq \text{const}(\eta, \nabla \eta, \|\nabla u_j\|_p, \|\nabla u_k\|_p) \cdot \sup_{\text{spt } \eta} |u_j - u_k|$$

for a constant which is controlled independent of j, k. Recalling $\|u_j - u_k\|_{L^\infty(\text{spt }\eta)} \xrightarrow[j,k\to\infty]{} 0$ we have proved our claim. □

We now apply Lemma 1.3.2 to a sequence which comes froms scaling (blowing up) a given minimizer. For $B > 0$ let

$$\mathcal{C}_B := H^{1,p}\text{–closure of all locally } \int_{B_1} |\nabla w|^p \, dx$$

$$\text{minimizing maps } u \in H^{1,p}(B_1, \overline{M}) \text{ s.t. } \int_{B_1} |\nabla u|^p \, dx \leq B$$

Lemma 1.3.3 *Suppose that $u \in \mathcal{C}_B$, $x_0 \in B_1$ and a sequence $r_i \downarrow 0$ are given. Then there is a subsequence r_i^* such that the scaled functions*

$$u_i(z) := u(x_0 + r_i^* z), \quad z \in B_1,$$

converge weakly in $H^{1,p}(B_1, \mathbb{R}^N)$ to some function $u_0 \in \mathcal{C}_B$. We have

(i) $\partial_r u_0 = 0$ *(radial independence)*

(ii) $\mathcal{H}^{n-p}(\text{Sing } u_0) = 0$

(iii) $u_i \to u_0$ *in $H^{1,p}_{\text{loc}}(B_1, \mathbb{R}^N)$ strongly and uniform on compact subsets of $B_1 - \text{Sing}(u_0)$.*

Remarks: In most applications u is just a minimizer but the class $\mathcal{C}_B$ is needed to carry out the general dimension reduction. In common notation u_0 is said to be a *tangent map* or a *blow-up limit* of u at x_0. It is an unsolved problem to prove the uniqueness of blow-up limits (i.e. independence of the choice of the subsequence r_i^*).

For minimizers u Lemma 1.3.3 is a direct consequence of Lemma 1.3.2, the radial independence of u_0 follows from the monotonicity formula. The general statement is a bit more technical.

Next we *prove Theorem B in case $n \leq p + 1$:* Recall that we have to show that $\text{Sing}(u)$ is discrete for any local minimizer $u \in H^{1,p}(B_1, \overline{M})$ of $\int_{B_1} |\nabla w|^p \, dx$ (Again we consider the easiest situation; more general functionals and/or domains with curvature are handled along similar lines.)

Let $\Sigma := \text{Sing}(u)$ and assume $\Sigma \ni y_i \to 0$, i.e. Σ is not discrete. Let

$$x_i := \frac{1}{2}\frac{y_i}{|y_i|}, \quad u_i(z) := u(2 \cdot |y_i| z)$$

for $z \in B_1$. After passing to subsequences and quoting Lemma 1.3.3 we have

1.3 Proofs of the Basic Theorems

$$x_i \to x, \quad |x| = \frac{1}{2}, \quad u_i \to u_0.$$

The limit point x is not in $\Sigma_0 := \text{Sing}(u_0)$. If x were in $\text{Sing}(u_0)$ then Σ_0 would contain the whole line $\{tx : 0 \leq t \leq 1\}$ due to the radial independence of u_0. On the other hand

$$\mathcal{H}^{n-p}(\Sigma_0) = 0 \underset{n-p \leq 1}{\Longrightarrow} \mathcal{H}^1(\Sigma_0) = 0$$

which is a contradiction. But $x \in \text{Reg } u_0$ implies[1]

$$\lim_{t \downarrow 0} t^{p-n} \int_{B_t(x)} |\nabla u_0|^p \, dz = 0$$

so that (recall $\nabla u_i \to \nabla u_0$ in L^p_{loc})

$$t^{p-n} \int_{B_t(x_i)} |\nabla u_i|^p \, dz < \varepsilon$$

for $i \gg 1$ and $t \ll 1$. Rewriting the inequality in terms of u we immediately see $y_i \in \text{Reg}(u)$ for $i \gg 1$. $\square$

We give a few comments concerning the *Proof of $C^{1,\varepsilon}$-regularity (Theorem C):* The basic ideas go back to Fusco & Hutchinson, I gave some minor technical modifications. Introduce the quantity (mean oscillation of ∇w)

$$\psi(w, B_r(x)) := |(\nabla w)_r|^{p-2} \fint_{B_r(x)} |\nabla w - (\nabla w)_r|^2 + \fint_{B_r(x)} |\nabla w - (\nabla w)_r|^p$$

and observe that if $v \in H^{1,p}(B_r(x), \mathbb{R}^N)$ is a solution of

$$-\partial_\alpha(|\nabla v|^{p-2} \partial_\alpha v) = 0$$

then

$$\psi(v, B_r(x)) \leq c \cdot \left(\frac{r}{R}\right)^\alpha \psi(v, B_R(x))$$

---

[1] Since the u_i are local minimizers we get

$$\int_{B_1} |\nabla u_i|^{p-2} \nabla u_i \cdot \nabla \Phi \, dz \leq a \cdot \int_{B_1} |\nabla u_i|^p \, |\Phi| \, dz$$

for any $\Phi \in H^{1,p} \cap L^\infty(B_1, \mathbb{R}^N)$ with compact support. Since we have strong local convergence $\nabla u_i \to \nabla u_0$ the above system extens to u_0 and using this it is not hard to show that at regular points the scaled energy has to vanish.

for suitable constants $c > 0$, $\alpha \in (0,1)$. Now, if u is a constrained local minimizer and x_0 is a regular point then a comparison argument similar to the one used in Theorem A implies

$$\psi(u, B_r(x)) \leq \text{const} \cdot r^\varepsilon$$

for all balls $B_r(x)$, x near x_0 and r small, hence $\nabla u \in C^{0,\varepsilon/p}$ near x_0. It should be noted that the modulus of Hölder continuity of ∇u can be estimated uniformly which is of importance for the

Proof of Theorem D, which is similar to the proof of Theorem B. We follow ideas of E. Giusti [49] who proved the same result in a different setting. Consider a minimizer u of $\int_{B_1} |\nabla u|^p \, dx$ in $H^{1,p}(B_1, \overline{M})$ with $0 \in \text{Sing}\, u$ and assume $n - 1 \leq p < n$ and

$$\lim_{i \to \infty} |x_i| \cdot |\nabla u(x_i)| = \infty$$

for a sequence $x_i \to 0$. We scale

$$u_i(z) := u(2 \cdot |x_i| \cdot z), \quad z \in B_1$$

and deduce from the monotonicity formula

$$\Phi(u_i, B_1) = \Phi(u, B_{2|x_i|}) \leq \Phi(u, B_R)$$

where $R > 0$ is chosen to satisfy $2|x_i| \leq R$. According to Lemma 1.3.3 we have

$$\begin{cases} u_i \rightharpoonup u_0 & \text{weakly in } H^{1,p}_{\text{loc}}(B_1, \mathbb{R}^N), \\ u_i \to u_0 & \text{strongly in } H^{1,p}_{\text{loc}}(B_1, \mathbb{R}^N), \\ u_i \to u_0 & \text{uniformly on compact subsets of } \text{Reg}(u_0). \end{cases}$$

Let $y_i := \frac{x_i}{(2|x_i|)} \to y$. $\mathcal{H}^1(\text{Sing}\, u_0) = 0$ gives $y \in \text{Reg}(u_0)$, hence

$$r^{p-n} \int_{B_r(y)} |\nabla u_i|^p < \varepsilon$$

for $i \gg 1$ and $r \ll 1$. Then by Theorem D ∇u_i is uniformly Hölder continuous on $B_{r/4}(y)$ and Arcela's Theorem implies

$$\nabla u_i \to \nabla u_0 \quad \text{uniformly on } B_{r/4}(y),$$

especially $\nabla u_i(y_i) \xrightarrow[i \to \infty]{} \nabla u_0(y)$ contradicting the choice of x_i. $\square$

The boundary regularity theorem E is proved in [32] and proceeds along the following lines:

1.3 Proofs of the Basic Theorems

- monotonicity of $r \mapsto r^{p-n} \int_{B_r(x) \cap \Omega} |\nabla u|^p \, dz$ for points $x \in \partial\Omega$

- a version of the Extension Theorem for half balls $B_r(x) \cap \Omega$

- partial Hölder continuity of u near $\partial\Omega$, i.e.:

$$\liminf_{r \downarrow 0} r^{p-n} \int_{B_r(x) \cap \Omega} |\nabla u|^p \, dz = 0 \quad \text{at } x \in \partial\Omega$$

implies $u \in C^{0,\alpha}(\overline{B_\rho(x) \cap \Omega})$

- dimension reduction for the singular set in $\partial\Omega$

The question of boundary regularity at $x_0 \in \partial\Omega$ can finally be reduced to the following problem: *if u_0 is a blow up limit $\mathbb{R}^n_+ \to \overline{M}$ s.t.*

$$u_0 \equiv const. \text{ on } \partial\mathbb{R}^n_+, \quad \text{Sing } u_0 = \{0\}$$

then u_0 is a constant map.

This fact can be proved using a special variation in the domain of definition which has been introduced by Hard & Lin [53]. □

Finally we study a situation for which $\mathrm{Sing}(u) = \emptyset$, i.e. we *prove Theorem F*: Consider a local minimizer $u \in H^{1,p}(B_1, \overline{M})$ of $\int_{B_1} |\nabla u|^p \, dx$ with M satisfying the hypothesis of Theorem F. We show $0 \notin \mathrm{Sing}(u)$ (so that by moving the origin $\mathrm{Sing}(u) = \emptyset$). Recall from Theorem 1.2.1

$$\int_{B_1} |\nabla u|^{p-2} \nabla u \cdot \nabla \zeta \, dx$$

$$= \int_{[u \in \partial M]} \Theta |\nabla u|^{p-2} (\mathcal{N}(u) \cdot \zeta) \left(\partial^2 d(u) (\partial_\alpha u, \partial_\alpha u) \right)_+ dx$$

for all $\zeta \in H^{1,p} \cap L^\infty(B_1, \mathbb{R}^N)$ with compact support. If M is star shaped w.r.t. to the point Q we insert $\zeta = \eta \cdot (u - Q)$ with $\eta \in C_0^1(B_1, [0,1])$, $\eta \geq 0$, and deduce from $\mathcal{N}(u) \cdot (u - Q) \leq 0$

$$\int_{B_1} |\nabla u|^{p-2} \nabla u \cdot \nabla(\eta(u - Q)) \, dx \leq 0.$$

If we consider a blow up sequence $u_i(z) := u(\lambda_i z)$ the above inequality is valid for u_i and also for the limit u_0. Consider $\eta(x) = \rho(|x|)$ with $\rho \in C_0^1(0,1)$, $\rho \geq 0$. Since $\partial_r u_0 \equiv 0$ we have $\partial_\alpha u_0(x) \cdot \partial_\alpha(\rho(|x|)) = 0$ so that

$$\int_{B_1} \rho(|x|)|\nabla u_0|^p \, dx = 0$$

and in conclusion $\nabla u_0 \equiv 0$ a.e. This implies

$$\lambda_i^{p-n} \int_{B_{\lambda_i}} |\nabla u|^p \, dx \xrightarrow[i \to \infty]{} 0,$$

hence $0 \in \text{Reg}(u)$. $\square$

1.4 A Survey on p-Harmonic Maps

Let
$\Omega\ =\ $ an open region of an n-dimensional Riemannian manifold
$Z\ =\ $ a k-dimensional submanifold of $\mathbb{R}^N$ with $\partial Z = \emptyset$

and $p > 1$ be given. The space $H^{1,p}(\Omega, Z)$ is defined in the standard way. Our results collected in section 1.2 especially include the (partial) regularity properties of minimizers of $\int_\Omega \|du\|^p \, d\text{vol}$ even when Z is a manifold with boundary.

Definition

a) $u \in H^{1,p}(\Omega, Z)$ is *weakly p-harmonic* iff u is a weak solution of the Euler equation corresponding to the energy $\int_\Omega \|du\|^p \, d\text{vol} =: \mathcal{E}_p(u, \Omega)$.

b) and a p-harmonic map is *p-stationary* if in addition u is critical w.r.t. reparametrisations of Ω.

Clearly minimizing maps are p-stationary but the converse does not hold. In this section we describe what is known about p-stationary mappings. We refer to [29], [11], [32], and [14]. Our first result is an existence theorem for small p-harmonic maps which extends a well known theorem of Hildebrandt & Kaul & Widman [61] on harmonic maps.

Theorem 1.4.1 *Let $g : \partial\Omega \to Z$ denote a smooth boundary function with values in a regular geodesic ball $\mathbb{B}_r(Q) \subset Z$. Then there exists $U \in H^{1,p}(\Omega, Z)$ as follows:*

(i) $U \in C^{0,\alpha}(\overline{\Omega}, Z)$ *for all $0 < \alpha < 1$ and $\in C^{1,\varepsilon}(\Omega, Z)$ for a certain $\varepsilon \in (0,1)$*

1.4 A Survey on p-Harmonic Maps

(ii) U is p-stationary

(iii) $U\big|_{\partial\Omega} = g$ and $U(\Omega) \subset \mathbb{B}_r(Q)$.

Remarks:

1. Let $\mathbb{B}_r(Q)$ denote a closed geodesic ball in Z, $C(Q)$ being the cut locus of Q. The ball is regular iff

   a) $\mathbb{B}_r(Q) \cap C(Q) = \emptyset$

   b) $r < \frac{\pi}{2\sqrt{\kappa}}$, $\kappa \geq 0$ being an upper bound for the sectional curvature of Z on $\mathbb{B}_r(Q)$.

   If $\operatorname{Sec}_Z \leq 0$ then any geodesic ball is regular. For example let $Z = S^N = \{u \in \mathbb{R}^{N+1} : |u| = 1\}$ and take $Q = (0, \ldots, 0, 1)$. Then the sets

   $$\{u \in Z : u^{N+1} \geq \delta\}$$

   are regular geodesic balls with center Q as long as $\delta > 0$. For $\delta = 0$ condition b) is violated.

2. From our construction it will follow that U is minimizing $\mathcal{E}_p(\cdot, \Omega)$ among all maps with values in $\mathbb{B}_r(Q)$ but U is not necessarily minimizing among all functions $\Omega \to Z$ with boundary values g.

The *Proof* combines our results on obstacle problems and a maximum principle: For $R > r$ but close to r $\mathbb{B}_R(Q)$ is also regular and we solve the obstacle problem

$$\mathcal{E}_p(\cdot, \Omega) \to \min \text{ in } H^{1,p}(\Omega, \mathbb{B}_R(Q)) \quad \text{for boundary values } g.$$

From section 1.1 we deduce $U \in C^{0,\alpha}(\overline{\Omega}, \mathbb{R}^N) \cap C^{1,\varepsilon}(\Omega, \mathbb{R}^N)$ and it remains to verify $U(\Omega) \subset \mathbb{B}_r(Q)$ which immediately will imply the stationarity of U. Let Ω be a flat domain with Euclidean metric and introduce normal coordinates $u = (u^1, \ldots, u^k)$ on $\mathbb{B}_R(Q)$ w.r.t. the center Q. Since the set $\mathbb{B}_R(Q)$ is geodesically star-shaped w.r.t. Q a similar reasoning as in the proof of Theorem F turns the Euler system for the obstacle problem into the differential inequality (compare [27])

$$\int_\Omega A(u, \nabla u) \left[\nabla u \cdot \nabla(\eta u) - \Gamma^\ell_{ij}(u)\, \partial_\alpha u^i\, \partial_\beta u^j\, u^\ell \eta \right] dx \leq 0 \qquad (1.4.1)$$

for all $\eta \in \overset{\circ}{H}{}^{1,p} \cap L^\infty(\Omega, \mathbb{R})$, $\eta \geq 0$. Here

$$\begin{cases} g_{ij} = \text{metric tensor of } Z, \ \Gamma^\ell_{ij} = \text{Christoffel symbols} \\ A(u, \nabla u) = \left(g_{ij}(u) \partial_\alpha u^i \partial_\alpha u^j\right)^{\frac{p}{2}-1}. \end{cases}$$

It should be noted that (1.4.1) can be proved by easier means if we use the fact that the side condition $|u| \leq R$ is convex by the way implying (1.4.1) as variational inequality.

We apply (1.4.1) with

$$\eta := \max(v - r^2, 0), \quad v := |u|^2,$$

which is admissible since $v \leq r^2$ on $\partial \Omega$. By a result of Hildebrandt [56] we have

$$|\nabla u|^2 - \Gamma^\ell_{ij}(u) \partial_\alpha u^i \partial_\alpha u^j u^\ell \leq 0$$

(here the curvature bound enters!), hence

$$\int_{\Omega \cap [|u| \geq r]} A(u, \nabla u) |\nabla v|^2 \, dx = 0$$

and $A(u, \nabla u) |\nabla \eta|^2 = 0$ on Ω. Since $|\nabla v| \leq 2R|\nabla u|$ and $A(u, \nabla u) \geq c \cdot |\nabla u|^{p-2}$ for some $c > 0$ we must have $\nabla \eta = 0$, hence $|u| \leq r$. □

For suitable boundary values it is therefore possible to construct *smooth* p-stationary mappings $\Omega \to Z$ with *small range*. This does not answer the general question of regularity of p-stationary mappings. We give some partial answers.

Theorem 1.4.2 *Assume that $u \in H^{1,p}(\Omega, Z)$ is weakly p-harmonic with $\text{Im}(U) \subset \mathbb{B}_R(Q)$ for a strongly regular ball. Then $\mathcal{H}^{n-p}(\text{Sing } U) = 0$ and $x \in \Omega$ is a regular point iff the scaled p-energy of U at x vanishes. If U is p-stationary then $\text{Sing } U = \emptyset$.*

Remarks:

1. A geodesic ball $\mathbb{B}_R(Q)$ is strongly regular if we require the curvature bound
$R < \frac{\pi}{4 \cdot \sqrt{\kappa}}$.

2. We conjecture $\text{Sing } U = \emptyset$ even under the weaker conditions that U is only p-harmonic and $R < \frac{\pi}{2\sqrt{\kappa}}$. This conjecture is motivated by results of Hildebrandt & Jost & Widman [60] obtained for harmonic maps ($p = 2$) where it is possible to use Green's function for the corresponding elliptic operator.

1.4 A Survey on p-Harmonic Maps

Proof of Theorem 1.4.2: If we introduce coordinates on $\mathbb{B}_R(Q)$ the representative u of U solves a degenerate elliptic system with right-hand-side of growth order p in ∇u and growth constant $a > 0$. If $\lambda > 0$ denotes a lower bound on the modulus of ellipticity then the assumption that $\mathbb{B}_R(Q)$ is strongly regular implies $a < \frac{\lambda}{2R}$. Hence we can prove Caccioppoli's inequality which gives $\nabla u \in L^{p+\varepsilon}_{\text{loc}}$ for some $\varepsilon > 0$. Now exactly the same arguments as in the proof of Theorem A gives $\mathcal{H}^{n-p}(\text{Sing } u) = 0$. For a more detailed exposition we refer to [29, Theorem 1.1]. We wish to emphasize the following fact:

> If the scaled p-energy is smaller than some constant ε then we have an apriori bound for the modulus of Hölder continuity on the corresponding ball in Ω.

Next we want to show $\text{Sing } U = \emptyset$ for stationary maps. For simplification we assume that Ω is a flat domain.

Lemma 1.4.1 (Monotonicity formula for stationary mappings) *Let $U: \mathbb{R}^n \supset \Omega \to Z \subset \mathbb{R}^N$ denote a p-stationary mapping. Then*

$$\rho^{p-n} \int_{B_\rho(x)} |\nabla U|^p \, dz - \sigma^{p-n} \int_{B_\sigma(x)} |\nabla U|^p \, dz$$

$$= p \cdot \int_{B_\rho(x) - B_\sigma(x)} |x - z|^{p-n} |\nabla U|^{p-2} |\partial_r U|^2 \, dz$$

for all balls $B_\sigma(x) \subset B_\rho(x) \subset \Omega$.

In section 1.3 we proved the formula for minimizers but here we have to modify the argument: The invariance of $\int_\Omega |\nabla U|^p \, dx$ against reparametrisations of Ω implies the 2nd Euler equation

$$0 = \int_\Omega \left(|\nabla U|^p \operatorname{div} X - p|\nabla U|^{p-2} \partial_\alpha U \cdot \partial_\beta U \, \partial_\alpha X^\beta \right) dx$$

for all $X \in C^1_0(\Omega, \mathbb{R}^n)$ (calculate $\frac{d}{dt/0} \int_\Omega |\nabla U_t|^p \, dx$ for $U_t(x) := U(x + t \cdot X(x))$) and we may proceed as in the paper [69].

We continue in the *Proof of Theorem 1.4.2:* It is enough to assume $0 \in \Omega$ and to show

$$r^{p-n} \int_{B_r(0)} |\nabla u|^p \, dx \xrightarrow[r \downarrow 0]{} 0$$

for the coordinate representative u of U w.r.t. geodesic coordinates centered at Q. Choose a sequence $r_\ell \downarrow 0$ and consider the scaled functions

$$U_\ell(z) := U(r_\ell \cdot z)$$

which are also weakly p-harmonic with range in the strongly regular ball. Let $\{u_\ell\}$ denote the sequence of coordinate representatives. Using $|u_\ell| \leq R$ we deduce a uniform Caccioppoli inequality, hence we have uniform bounds in the spaces $H^{1,p}(B_t(0))$ for any $0 < t < 1$ which implies for subsequences

$$U_\ell \to U_0 \quad (u_\ell \to u_0) \begin{cases} \text{weakly in } H^{1,p}_{\text{loc}}(B_1) \\ \text{strongly in } L^p_{\text{loc}}(B_1) \\ \text{pointwise a.e.} \end{cases}$$

Suppose now that we already know

$$\nabla U_\ell \to \nabla U_0 \quad \text{strongly in } L^p_{\text{loc}}(B_1). \tag{1.4.2}$$

Then the monotonicity being valid for each U_ℓ extends to U_0 and implies $\partial_r U_0 \equiv 0$ (this is an easy calculation using Lemma 1.4.1 for U_ℓ). Moreover (1.4.2) also shows that U_0 is p-harmonic, i.e.

$$0 = \int_{B_1} A(u_0, \nabla u_0) \left(\nabla u_0 \cdot \nabla \psi - \Gamma^\ell_{ij}(u_0) \nabla u_0^i \cdot \nabla u_0^j \psi^\ell \right) dx \tag{1.4.3}$$

for all test vectors. Here we have used the notations from the proof of Theorem 1.4.1. Inserting $\psi(x) = \eta(|x|) \cdot u_0(x)$ with $\eta \in C^1_0(0,1)$ and using $\partial_r u_0 = 0$ we arrive at

$$0 = \int_{B_1} A(u_0, \nabla u_0) \, \eta \cdot [|\nabla u_0|^2 - \Gamma^\ell_{ij}(u_0) \cdot \nabla u_0^i \cdot \nabla u_0^j \, u_0^\ell] \, dx \, .$$

As in the proof of Theorem 1.4.1 we deduce $\nabla u_0 = 0$ which implies

$$r_\ell^{p-n} \int_{B_{r_\ell}(0)} |\nabla u|^p \, dx \xrightarrow[\ell \to \infty]{} 0$$

so that $0 \in \text{Reg}(u)$.

Our proof will be completed by showing strong convergence (1.4.2). We use essentially the same ideas as in the proof of Lemma 1.3.2.

1st *step:* By combining Caccioppli's inequality with the partial regularity criterion for weakly p-harmonic mappings with range in a strongly regular ball we obtain

$$\left| \begin{array}{l} \text{If } \fint_{B_r(x)} |u - (u)_r|^p \, dz < \varepsilon^p \text{ for some ball } B_r(x) \subset B_1 \text{ then } u \in \\ C^{0,\alpha}(B_{r/2}(x)) \text{ uniformly (i.e. Hölder exponent and Hölder constant do not depend on } u). \text{ Here } \varepsilon \text{ is a small absolute constant.} \end{array} \right.$$

1.4 A Survey on p-Harmonic Maps

2nd step: Consider $x \in B_1$ s.t.

$$\fint_{B_r(x)} |u_0 - (u_0)_r|^p \, dx < \varepsilon^p .$$

Then $\fint_{B_r(x)} |u_\ell - (u_\ell)_r|^p \, dx < \varepsilon^p$ and by step 1 (at least for $\ell \gg 1$, recall $u_\ell \to u_0$ in L^p_{loc})

$$u_\ell \in C^{0,\alpha}(B_{r/2}(x)), \quad u_\ell \to u_0 \text{ uniformly}$$

on $B_{r/2}(x)$. As in Lemma 1.3.2 we infer

$$\nabla u_\ell \to \nabla u_0 \quad \text{in} \quad L^p_{\text{loc}}(\text{Reg } u_0) .$$

On the other hand

$$\text{Sing } u_0 \subset \left\{ x \in B_1 : \liminf_{r \downarrow 0} \fint_{B_r(x)} |u_0 - (u_0)_r|^p \, dx > 0 \right\}$$

so that $\mathcal{H}^{n-p}(\text{Sing } u_0) = 0$ and the contribution to the energy is as small as we want (compare Lemma 1.3.2). □

Without imposing any smallness conditions on the range of a p-harmonic map U nothing is known about the regularity properties of these functions. On the other hand there is a great number of papers being concerned with "*removable singularities*" in different settings, e.g. [64] for harmonic maps or [50]. In [11] we obtained the following

Theorem 1.4.3 *Let B_1 denote the unit ball in $\mathbb{R}^n$ and suppose that $u \in C^1(B_1 - \{0\}, \mathbb{R}^N) \cap H^{1,p}(B_1, \mathbb{R}^N)$ is weakly p-harmonic with range in the manifold Z. Then there is a constant $\varepsilon > 0$ depending on n, p, N and the geometry of Z such that $u \in C^{1,\gamma}(B_1)$ for some $\gamma \in (0,1)$ provided*

$$\int_{B_1} |\nabla u|^p \, dx < \varepsilon .$$

Remarks:

1. By scaling we deduce that an isolated singular point x_0 is removable if

$$\Phi(u, B_r(x_0)) = r^{p-n} \int_{B_r(x_0)} |\nabla u|^p \, dx < \varepsilon$$

    for some ball in the domain of definition.

2. Clearly Theorem 1.4.3 extends to the case when the domain of definition is a Riemannian manifold.

3. In the limit case $p = n$ the conformal invariance of the n-energy clearly implies removability of the possible singularity if $\int_{B_1} |\nabla u|^n \, dx < \infty$.

4. For simplicity the target Z is assumed to be compact but actually a curvature bound is sufficient.

5. Coron and Gulliver [6] showed that $(p < n) u_* : B_1 \to \partial B_1$, $x \mapsto x \cdot |x|^{-1}$, is p-energy minimizing in $\{v \in H^{1,p}(B_1, \partial B_1) : v(x) = x \text{ on } \partial B_1\}$. Therefore, u_* is p-stationary with finite p-energy and $0 \in \operatorname{Sing} u_*$. Hence our small energy assumption is really necessary. Unfortunately we have no explicit bound for ε but we conjecture $\varepsilon = \int_{B_1} |\nabla u_*|^p \, dx$ at least for $Z = \partial B_1$.

6. The example in 5. shows that linear growth

$$\limsup_{x \to 0} |x| \cdot |\nabla u(x)| < \infty \tag{1.4.4}$$

is not sufficient for regularity at 0. But we can prove

Theorem 1.4.4 *Suppose that $u \in C^1(B_1 - \{0\}, Z)$ is p-harmonic with (1.4.4) from above and the small range condition $u(B_1 - \{0\}) \subset \mathbb{B}_r(Q)$ for a regular geodesic ball $\mathbb{B}_r(Q)$. Then $u \in C^{1,\gamma}(B_1)$ for some $\gamma \in (0,1)$.*

Note that our example u_* just violates the small range condition. For minimizing maps and under the restriction $n - 1 \leq p < n$ linear growth at the possible singular points was proved in section 1.3. From this point of view it is rather natural to impose (1.4.4),

7. In recent times many people have started to investigate weakly harmonic maps $u : B^n \to \partial B^n$ (or some S^N) without imposing any apriori assumptions on the behaviour of u. We mention the result of Evans [22] who proved $\mathcal{H}^{n-2}(\operatorname{Sing} u) = 0$ for any stationary harmonic map $u \in H^{1,2}(B^n, S^N)$. A similar theorem can be shown to hold for p-stationary mappings.

Outline of the proof of Theorem 1.4.3: The crucial part is an apriori estimate for p-harmonic maps of class C^1.

Lemma 1.4.2 *Suppose that $u \in C^1(B_r(x), Z)$ is p-harmonic. Then $\Phi(u, B_r(x)) < \varepsilon_1$ implies*

1.4 A Survey on p-Harmonic Maps

$$\sup_{B_{r/2}(x)} |\nabla u|^p \leq C_0 \fint_{B_r(x)} |\nabla u|^p \, dz \, .$$

Here ε_1, C_0 are positive constants independent of u.

Proof:

1. By using difference quotient technique we prove
$$V := |\nabla u|^{\frac{p}{2}-1} \nabla u \in H^{1,2}_{\text{loc}}(B_r(x)) \, .$$

2. We use 1. to derive a differential inequality for $|\nabla u|^p$: There exists an absolute constant $K > 0$ such that
$$\int \left(a_{\alpha\beta}(\cdot, \nabla u) \partial_\beta (|\nabla u|^p) \, \partial_\alpha \varphi - K |\nabla u|^{p+2} \varphi \right) dz \leq 0 \qquad (1.4.5)$$
for all $\varphi \geq 0$ with compact support in $B_r(x) \cap [\nabla u \neq 0]$.

3. We extend (1.4.5) to all $\varphi \geq 0$ with support in $B_r(x)$. Here $a_{\alpha\beta}(\cdot, \nabla u)$ denote bounded measurable and elliptic coefficients.

4. The monotonicity formula holds for p-harmonic maps which are C^1 up to an isolated singular point. (Recall that in Lemma 1.4.1 we required stationarity.)

5. By applying a maximum principle to inequality (1.4.5) the claim of Lemma 1.4.2 follows.

□

The next step towards the proof of Theorem 1.4.3 is a discrete Morrey condition:

Lemma 1.4.3 *There are constants ε_0 and σ such that for any p-harmonic maps $u \in C^1(B_1 - \{0\}, Z) \cap H^{1,p}(B_1)$ with $\int_{B_1} |\nabla u|^p \, dx \leq \varepsilon_0$ we have*
$$\Phi(u, B_\sigma) \leq \frac{1}{2} \Phi(u, B_1) \, .$$

Proof: Argueing by contradiction we find a sequence of p-harmonic maps $u_i \in C^1(B_1 - \{0\}, Z)$ such that
$$\Phi(u_i, B_1) \leq \frac{1}{i} \to 0 \, ,$$
$$\Phi(u_i, B_\sigma) \geq \frac{1}{2} \Phi(u_i, B_1) \, .$$

We pass to the normalized sequence
$$v_i := \Phi(u_i, B_1)^{-1/p} \left(u_i - (u_i)_1\right)$$
and get (after passing to subsequences) $v_i \rightharpoonup v_\infty$ in $H^{1,p}$ and
$$\lim_{i \to \infty} \int_{B_1} |\nabla v_i|^{p-2} \nabla v_i \nabla \varphi \, dx = 0$$
for all $\varphi \in C_0^1(B_1)^N$. Using the apriori bound it is easy to show $\nabla v_i \to \nabla v_\infty$ in $L^p(B_{1/2} - B_r)$ for any $r > 0$. But on B_r all gradients have uniformly small L^p-norm so that v_∞ solves $\partial_\alpha(|\nabla v_\infty|^{p-2} \partial_\alpha v_\infty) = 0$ on B_1 which leads to a contradiction. This proves Lemma 1.4.3 and by the way Theorem 1.4.3. □

Proof of Theorem 1.4.4: We have to show $\liminf_{\rho \downarrow 0} \rho^{p-n} \int_{B_\rho(0)} |\nabla u|^p \, dx = 0$. To this purpose we choose a sequence $\lambda_i \downarrow 0$ and define
$$u_i(z) := u(\lambda_i z), \quad |z| < 1.$$
From our assumption
$$|x| \cdot |\nabla u(x)| \leq K < \infty, \quad x \neq 0,$$
we obtain $|\nabla u_i(x)| \leq K \cdot |x|^{-1}$ so that $u_i \rightharpoonup u_0$ locally uniformly on $B_1 - \{0\}$. It is then easy to prove that
$$\int_{B_1 - B_r} |\nabla u_i - \nabla u_j|^p \, dx \xrightarrow[i,j \to \infty]{} 0$$
for any positive r, on the other hand for any $\delta > 0$ we find $r > 0$ such that
$$\int_{B_r} |\nabla u_i|^p \, dx \leq \delta \quad \forall i \gg 1$$
so that $u_i \to u_0$ strongly in $H^{1,p}(B_1)$. Hence u_0 satisfies the monotonicity formula which implies $\partial_r u_0 = 0$. Then, using the Euler equation for u_0 we arrive at $u_0 = 0$ which gives the result. □

A celebrated theorem of Eells & Sampson [21] applies the *heat flow method* to obtain the *existence of harmonic maps in homotopy classes of mappings* $M \to N$ provided $\text{Sec}^N \leq 0$. (The sectional curvature of N at y along the section $S = a$ 2-dimensional subspace of $T_y N$ corresponds to the Gauss curvature of $\exp_y(S \cap$ a neighborhood of 0 in $T_y N$) and we say $\text{Sec}^N \leq 0$ if this scalar quantity is ≤ 0 for all y and S.) When trying to extend their result to the p-case one is faced with the difficulty that now the heat (flow) operator $\mathcal{L}(u) = \frac{\partial}{\partial t} u - \Delta u$ has to be replaced by $\tilde{\mathcal{L}}(u) = \frac{\partial}{\partial t} u - \partial_\alpha(|\nabla u|^{p-2} \partial_\alpha u)$ where no comparible theory is available. Before giving an outline of an alternative approach which also works for the p-energy functional let us fix the precise assumptions:

1.4 A Survey on p-Harmonic Maps

M^m, N^n = compact Riemannian manifolds without boundary;

$\qquad\qquad N^n \hookrightarrow \mathbb{R}^k$ for some k

$H^{1,p}(\Omega, N) := \{v \in H^{1,p}(\Omega, \mathbb{R}^k); v(x) \in N \text{ a.e.}\}$

(Here Ω is an arbitrary open subset of M.)

$E_p(u, \Omega) := \int_\Omega |du|^p \, dv \qquad p\text{-energy}$

$E_p^\delta(u, \Omega) := \int_\Omega (\delta + |du|^2)^{p/2} \, dv \qquad (\delta, p)\text{-energy} \quad (\text{for } \delta > 0).$

For $\varphi \in C^0(M, N)$ we let

$[\varphi] := \left\{ \psi \in C^0(M, N) : \begin{array}{l} \exists H \in C^0(M \times [0,1], N) \text{ s.t.} \\ H(\cdot, 0) = \varphi, \; H(\cdot, 1) = \psi \end{array} \right\}$

(homotopy class of φ)

Theorem 1.4.5 *If* $\operatorname{Sec}^N \leq 0$ *then* $E_p(\cdot, M)$ *attains its minimum in* $[\varphi]$, *more precisely: There exists a p-harmonic map* u *of class* C^1 *such that* $u \in [\varphi]$ *and* $E_p(u, M) \leq E_p(w, M)$ *for all* $w \in [\varphi]$.

(see [14])

Remark: The method extends to the case $\partial M \neq \emptyset$ with an additional Dirichlet boundary condition.

Outline of the proof: One major advantage of introducing the less degenerate E_p^δ-functional is the better regularity of the minimizers.

Theorem 1.4.6 *If* $u \in C^{1,\alpha}(M, N)$ *is weakly* (δ, p)-*harmonic then* u *is smooth.*

Proof: Show that any derivative of u is a solution of a linear elliptic system with Hölder coefficients, hence $du \in C^{1,\alpha}$; apply standard potential theory.

Similar to Theorem 1.4.1 we prove

Lemma 1.4.4 *Let* Ω *denote an open region in* M *with* $\partial\Omega \neq \emptyset$ *and consider a smooth boundary function* $u_0 : \partial\Omega \to N$ *with* $u_0(\partial\Omega) \subset \mathbb{B}_r(q)$ *for a regular geodesic ball. Then there exists* $u \in H^{1,p}(\Omega, \mathbb{B}_r(q))$ *as follows:*

(i) $\quad u \in C^\infty(\Omega)$

(ii) $\quad u$ *is* (δ, p)-*harmonic*

(iii) $u\big|_{\partial\Omega} = u_0$ and $u \in C^{0,\alpha}(\overline{\Omega})$ for any $\alpha \in (0,1)$.

$\square$

Lemma 1.4.5 *Assume* $\mathrm{Sec}^N \leq 0$ *and let* $\delta > 0$ *and* $\varphi \in C^0(M,N) \cap H^{1,p}$ *be given. Then there is a smooth* (δ,p)-*harmonic map* $u \in [\varphi]$ *such that* $E_p^\delta(u,M) = \inf_{[\varphi]} E_p^\delta(\cdot,M)$.

Proof: Let
$$E_0 := \inf_{[\varphi]} E_p^\delta(\cdot,M)$$
and consider a minimizing sequence $\{u_i\}$. W.l.o.g. we may assume $u_i \in C^\infty(M,N)$ and after passing to subsequences we have

$$u_i \rightarrow: u_0 \begin{cases} \text{weakly in } H^{1,p} \\ \text{strongly in } L^p \\ \text{a.e. on } M \end{cases}.$$

Fix $x_0 \in M$ and consider a geodesic ball in M with center x_0 and radius R. On account of $\mathrm{Sec}^N \leq 0$ any geodesic ball $\mathbb{B}$ is regular in N, thus by Lemma 1.4.4 there are (δ,p)-harmonic maps $h_i \in C^\infty(B_R(x_0),N)$ with $h_i = u_i$ on $\partial B_R(x_0)$. Using an apriori bound similar to Lemma 1.4.2 (now for (δ,p)-harmonic maps, the smallness of energy can be replaced by $\mathrm{Sec}^N \leq 0$) we see $h_i \rightarrow: h_0$ uniformly on compact subsets of $B_R(x_0)$ and h_0 clearly is (δ,p)-harmonic.

We expect $u_0 = h_0$ near x_0: for $Q_1, Q_2 \in N$ let $c(\cdot, Q_1, Q_2) : [0,1] \to N$ denote the unique geodesic arc from Q_1 to Q_2 and consider

$$H_i(x,t) := c(t, h_i(x), u_i(x)), \quad x \in B_R(x_0).$$

The claim then follows by discussing $\frac{d^2}{dt^2} E_p^\delta(H_i(t,\cdot), B_R(x_0))$.

Now $u_0 = h_0$ near x_0 implies by the arbitrariness of x_0 that u_0 is a smooth (p,δ)-harmonic map with

$$E_0 = E_p^\delta(u_0, M).$$

It remains to show $u_0 \in [\varphi]$. Let $B_{r/2}(x_i)$ denote a finite covering of M. On $B_r(x_1)$ we replace u_i by h_i, i.e. we define

$$u_i^1 := \begin{cases} u_i \text{ on } M - B_r(x_1) \\ h_i \text{ on } B_r(x_1) \end{cases}$$

1.4 A Survey on p-Harmonic Maps

u_i^1 is also a minimizing sequence with $u_i^1 \in [\varphi]$ and uniformly convergent to u_0 on $B_{r/2}(x_1)$. Repeating the argument on $B_r(x_2)$ we pass to another minimizing sequence u_i^2 in $[\varphi]$ s.t. $u_i^2 \to u_0$ uniformly on $B_{r/2}(x_1) \cup B_{r/2}(x_2)$, etc. Finally we have a minimizing sequence $w_i \in [\varphi]$ s.t. $w_i \to u_0$ uniformly on M, thus $u_0 \in [\varphi]$. □

We are now in the position to prove Theorem 1.4.5: Let h_δ be a minimizer of $E_p^\delta(\cdot, M)$ in $[\varphi]$. We have the apriori estimate

$$\|dh_\delta\|_{L^\infty(M)}^p \leq c \cdot \int_M \left(\delta + |dh_\delta|^2\right)^{p/2} dv$$

$$\leq c \cdot \int_M \left(1 + |d\varphi|^2\right)^{p/2} dv$$

so that $h_\delta \xrightarrow[\delta \downarrow 0]{} h$ uniformly for a Lipschitz map $h \in [\varphi]$. For any $w \in [\varphi]$ we have

$$E_p(h_\delta, M) \leq E_p^\delta(h_\delta, M) \leq E_p^\delta(w, M) \xrightarrow[\delta \downarrow 0]{} E_p(w, M),$$

on the other hand

$$E_p(h, M) \leq \liminf_{\delta \downarrow 0} E_p(h_\delta, M)$$

so that

$$E_p(h, M) \leq E_p(w, M).$$

h is p-harmonic: if π denotes a smooth retraction onto N then $\pi(h+t\psi) \in [\varphi]$ for $|t| << 1$ and $\psi \in C^1(M, \mathbb{R}^k)$. Hence $0 = \frac{d}{dt/0} E_p\big(\pi(t+t\psi)\big)$ and the claim follows. □

We close this section with some comments on *parametrized H-surfaces in codimension* 1 which are closely related to m-harmonic maps on m-dimensional domains. For details we refer to [12].
Let

$$\begin{cases} \Omega \subset \mathbb{R}^m & \text{denote an open region,} \\ H : \mathbb{R}^{m+1} \to \mathbb{R} & \text{a bounded Lipschitz function} \end{cases}$$

and consider a boundary function $u_0 \in H^{1,m}(\Omega, \mathbb{R}^{m+1})$. We are then looking for a solution of

$$\begin{cases} \partial_\alpha(|\nabla u|^{m-2} \partial_\alpha u) = m^{m/2} H(u) J(u) & \text{on } \Omega \\ u = u_0 & \text{on } \partial\Omega \end{cases} \tag{1.4.6}$$

$$J(u) := \sum_{i=1}^{m+1} (-1)^{i+1} \det \nabla \widehat{u^i} \cdot e_i,$$

$$\widehat{u^i} = (u^1 \ldots u^{i-1} u^{i+1} \ldots u^{m+1}), \quad e_i = (0 \ldots 1 \ldots 0).$$

In two dimensions (1.4.6) is just the Dirichlet problem for surfaces of prescribed mean curvature H treated by Hildebrandt, Heinz, Jäger, Kaul and many authors in the seventies, for $m \geq 3$ it has a similar geometric meaning but the two-dimensional methods do not apply.

Theorem 1.4.7 *If $\|u_0\|_{L^\infty} < \rho$ for some $\rho > 0$ and if we have the curvature bounds*

$$|H(z)| < \frac{m+1}{m} \rho^{-1}, \quad |z| \leq \rho$$

$$\sup_{|z|=\rho} |H(z)| \leq \rho^{-1}$$

then (1.4.6) admits a solution $u \in H^{1,m}(\Omega, \mathbb{R}^{m+1})$, $\|u\|_\infty < \rho$.

Proof: Minimize

$$u \mapsto \int_\Omega \left(|\nabla u|^m + m^{1+\frac{m}{2}} Q(u) \cdot J(u) \right) dx$$

in $H^{1,m}(\Omega, \mathbb{R}^{m+1}) \cap \{v : v = u_0 \text{ on } \partial\Omega, |v| \leq \rho \text{ a.e.}\}$; here Q is chosen to satisfy $\operatorname{div} Q = H$. The solution of Theorem 1.4.7 is of class $C^{1,\mu}(\Omega)$ for some $0 < \mu < 1$.

We give some ideas of how to obtain regularity of minimizers. Following the arguments used in [18] we show that any bounded weak solution $u \in H^{1,m}(\Omega, \mathbb{R}^m)$ of (1.4.6) has Hölder continuous first derivatives which in particular includes the result for minimizers. Our main tool is a version of the isoperimetric inequality in codimension 1 which needs some preparations:

Let $B \subset \mathbb{R}^m$ denote the unit ball. To $u, v \in H^{1,m}(B, \mathbb{R}^{m+1})$ we associate the currents (see section 2.2)

$$J_u \in \mathcal{D}_m(\mathbb{R}^{m+1}), \quad I_{u,v} \in \mathcal{D}_{m+1}(\mathbb{R}^{m+1}),$$

$$J_u(\beta) := \int_B \langle \beta \circ u, D_1 u \wedge \ldots \wedge D_m u \rangle \, dx = \int_B u^\# \beta, \quad \beta \in \mathcal{D}^m(\mathbb{R}^{m+1}),$$

$$I_{u,v}(\gamma) := \int_B \int_0^1 \langle \gamma(h(t,x)), (u(x)-v(x)) \wedge D_1 h(t,x) \wedge \ldots \wedge D_m h(t,x) \rangle \, dt \, dx$$

$$= \int_{[0,1] \times B} h^\# \gamma, \quad \gamma \in \mathcal{D}^{m+1}(\mathbb{R}^{m+1}),$$

1.4 A Survey on p-Harmonic Maps

where $h(t,x) := tu(x) + (1-t)v(x)$. Then we have (compare [78])

Lemma 1.4.6 *Suppose that $u,v \in H^{1,m}(B,\mathbb{R}^{m+1})$ and $u-v \in \overset{\circ}{H}{}^{1,m}(B,\mathbb{R}^{m+1})$. Then the following statements are true:*

(i) $\partial I_{u,v} = J_u - J_v$.

(Here ∂ denotes the boundary operator for currents (section 2.2).)

(ii) There exists a (unique) integer valued function $i_{u,v} \in L^1(\mathbb{R}^{m+1}) \cap L^{\frac{m+1}{m}}(\mathbb{R}^{m+1})$ representing $I_{u,v}$, i.e.

$$I_{u,v}(\gamma) = \int_{\mathbb{R}^{m+1}} i_{u,v}\,\gamma\,.$$

(iii)
$$\int_{\mathbb{R}^{m+1}} |i_{u,v}|\,dz \leq c_1(m)\,[A(u)+A(v)]^{\frac{m+1}{m}},$$

$$c_1(m) := (m+1)^{-1-1/m}\,\alpha(m)^{-1/m}, \quad \alpha(m) = \mathcal{L}^m(B),$$

$$A(u) := \int_B |J(u)|\,dx\,.$$

Remark: In [78] Lemma 1.4.6 is demonstrated only in the twodimensional case $m=2$ but it is not hard to prove appropriate versions for all $m \geq 3$.

For $u \in H^{1,m}(B;\mathbb{R}^{m+1}) \cap L^\infty$ let us set

$$V(u) = \frac{1}{m+1}\int_B u \cdot J(u)\,dx$$

which is just the volume of the cone over the surface $u(B)$. Observing the relation

$$V(u) - V(v) = \int_{\mathbb{R}^{m+1}} i_{u,v}\,dz$$

we can reformulate (iii) of Lemma 1.4.6 in the following way.

Lemma 1.4.7 *Suppose that the functions $u,v \in H^{1,m}(B,\mathbb{R}^{m+1}) \cap L^\infty$ coincide on ∂B. Then we have the estimate*

$$|V(u) - V(v)| \leq c_1(m)\,[A(u)+A(v)]^{1+1/m}$$
$$\leq c_1(m)\,m^{-m/2}[E(u,B)+E(v,B)]^{1+1/m},$$

$$E(u,B) := \int_B |\nabla u|^m\,dx,$$

where $c_1(m)$ is defined in the previous lemma.

We next apply Lemma 1.4.7 to prove the following important inequality:

Lemma 1.4.8 *For $\varphi \in \overset{\circ}{H}{}^{1,m}(B,\mathbb{R}^{m+1}) \cap L^\infty$ and $u \in H^{1,m}(B,\mathbb{R}^{m+1}) \cap L^\infty$ we have the estimate*

$$\left| \int_B \det(\varphi, D_1 u, \ldots, D_m u) dx \right| = \left| \int_B \varphi \cdot J(u) \, dx \right| \leq c_2(m) \, E(\varphi, B)^{1/m} \, E(u, B). \tag{1.4.7}$$

In order to prove (1.4.7) let us define the functionals

$$\psi(v^1, \ldots v^{m+1}) := \int_B \det \begin{pmatrix} v^1 & \cdots & v^{m+1} \\ \nabla v^1 & \cdots & \nabla v^{m+1} \end{pmatrix} dx ,$$

$$L_u(\varphi) := \sum_{i=1}^{m+1} \psi(u^1, \ldots, u^{i-1}, \varphi^i, u^{i+1}, \ldots, u^{m+1}) .$$

We apply Lemma 1.4.7 to

$$\tilde{u} := \left(\frac{u^1}{\|u^1\|}, \ldots, \frac{u^{i-1}}{\|u^{i-1}\|}, \frac{\varphi^i}{\|\varphi^i\|}, \frac{u^{i+1}}{\|u^{i+1}\|}, \ldots, \frac{u^{m+1}}{\|u^{m+1}\|} \right)$$

$$\tilde{v} := \left(\tilde{u}^1, \ldots, \tilde{u}^{i-1}, \frac{-\varphi^i}{\|\varphi^i\|}, \tilde{u}^{i+1}, \ldots, \tilde{u}^{m+1} \right)$$

(assuming that all the occurring seminorms $\|w\| = E(w,B)^{1/m}$ are different from zero – otherwise (1.4.7) is trivial) and end up with

$$|\psi(u^1, \ldots u^{i-1}, \varphi^i, u^{i+1}, \ldots, u^{m+1})|$$
$$\leq c_3(m) \|u^1\| \ldots \|u^{i-1}\| \cdot \|\varphi^i\| \cdot \|u^{i+1}\| \ldots \|u^{m+1}\| .$$

Taking the sum from 1 to $m+1$ we see that $|L_u(\varphi)|$ is bounded by the righthand side of (1.4.7). Finally we observe ([68], 4.4.6)

$$L_u(\varphi) = (m+1) \int_B \det(\varphi, D_1 u, \ldots, D_m u) \, dx$$

which gives the desired result.

Let $u \in H^{1,m}(\Omega, \mathbb{R}^{m+1})$ denote a bounded solution of equation (1.4.6). We fix a ball $B_R(x_0) \subset \Omega$ and calculate the solution v of $\partial_\alpha(|\nabla v|^{m-2} \partial_\alpha v) = 0$ on $B_R(x_0)$ for boundary values u. From [80] we infer ($c_4, c_5 \ldots$ denoting absolute constants)

$$\int_{B_r(x_0)} |\nabla v|^m \, dx \leq c_4 \left(\frac{r}{R}\right)^m \int_{B_R(x_0)} |\nabla u|^m \, dx \tag{1.4.8}$$

1.4 A Survey on p-Harmonic Maps

for all $0 < r \leq R$. On the other hand we may insert $u - v$ as test-vector in (1.4.6) and use (1.4.7) to obtain

$$\int_{B_R(x_0)} |\nabla u - \nabla v|^m dx$$

$$\leq c_5 \int_{B_R(x_0)} \{|\nabla u|^{m-2} \nabla u - |\nabla v|^{m-2} \nabla v\} \cdot (\nabla u - \nabla v) \, dx$$

$$= c_6 \int_{B_R(x_0)} \det \left(H(u)(u - v), D_1 u, \ldots, D_m u \right) dx$$

$$\leq c_7 \Big\{ \int_{B_R(x_0)} |\nabla (H(u)(u - v))|^m dx \Big\}^{1/m} \int_{B_R(x_0)} |\nabla u|^m \, dx \, .$$

Observing

$$|\nabla (H \circ u)| \leq \mathrm{Lip}(H) |\nabla u|$$

and

$$E\big(v, B_R(x_0)\big) \leq E\big(u, B_R(x_0)\big)$$

we get in combination with (1.4.8)

$$\int_{B_r(x_0)} |\nabla u|^m \, dx \leq c_8 \Big\{ \Big(\frac{r}{R}\Big)^m + E\big(u, B_R(x_0)\big)^{1/m} \Big\} \int_{B_R(x_0)} |\nabla u|^m dx \, .$$

From this inequality we easily deduce $u \in C^{0,\alpha}$ for any $0 < \alpha < 1$ (observe $E\big(u, B_R(x_0)\big)^{1/m} \to 0$ as $R \downarrow 0$) and higher regularity follows along the lines of Theorem C. $\square$

Applying the methods developed for p-harmonic maps to problem (1.4.6) we obtained various regularity theorems for arbitrary weak solutions of (1.4.6). Among other things we mention a result on *removable isolated singularities* which follows from apriori gradient bounds and monotonicity formulas. Moreover, one can prove the following:

> If u_k is a sequence of regular solutions to (1.4.6) with bounded m-energy then there are finitely many points $x_1, \ldots, x_L \in \Omega$ such that $u_k \to: u$ strongly in $H^{1,m}_{\mathrm{loc}}(\Omega - \{x_1, \ldots, x_L\})$. u is a weak solution of (1.4.6) on Ω and of class $C^{1,\alpha}$ outside the points $x_1, \ldots, x_L$.

But unfortunately the parametric approach to H-surfaces has its full geometric meaning only in case $m = 2$. So we will treat H-manifolds more detailed in a Geometric Measure Theory setting.

Appendix

For completeness we would like to describe the blow-up behaviour of sequences of p-harmonic mappings (compare [38]).

Theorem 1.4.8 *Let Ω denote an open bounded set in $\mathbb{R}^m$, $m \geq 2$, and consider a sequence $\{u_n\}$ in $H^{1,p}(\Omega, \mathbb{R}^k)$ such that $u_n \rightharpoonup u$ weakly in this space and*

$$\int_\Omega |\nabla u_n|^{p-2} \nabla u_n \cdot \nabla \varphi \, dx \leq c_n \|\varphi\|_{L^\infty(\Omega)} \qquad (1.4.9)$$

for all $\varphi \in \overset{\circ}{H}{}^{1,p}(\Omega, \mathbb{R}^k) \cap L^\infty$ with $\lim_{n \to \infty} c_n = 0$. Then u is a p-harmonic mapping $\Omega \to \mathbb{R}^k$, i.e. $\partial_\alpha(|\nabla u|^{p-2}\partial_\alpha u) = 0$ in the weak sense. Here $p \in (1, \infty)$ denotes an arbitrary real number.

Going through the arguments outlined in Evan's paper [22] we obtain as a

Corollary: *Assume that $u \in H^{1,p}(\Omega, S^{k-1})$ is a weakly p-harmonic map from the domain Ω into the sphere S^{k-1} which in addition satisfies the monotonicity inequality*

$$R^{p-m} \int_{B_R(x)} |\nabla u|^p dy \geq r^{p-m} \int_{B_r(x)} |\nabla u|^p dy$$

for all balls $B_r(x) \subset B_R(x) \subset \Omega$. Then $u \in C^1(\Omega - \Sigma)$ for a relatively closed set $\Sigma \subset \Omega$ such that $\mathcal{H}^{m-p}(\Sigma) = 0$.

Remarks:

1. As mentioned in section 1.1 Theorem 1.4.8 implies partial regularity for p-energy minimizing maps.

2. If (1.4.9) is replaced by the weaker condition

$$\int_\Omega |\nabla u_n|^{p-2} \nabla u_n \cdot \nabla \varphi \, dx \leq c\|\varphi\|_{L^\infty(\Omega)} \qquad (1.4.10)$$

   for some positive constant c then we have

$$\left.\begin{array}{l} \nabla u_n(x) \to \nabla u(x) \quad \text{a.e.} \\ |\nabla u_n|^{p-2}\nabla u_n \rightharpoonup |\nabla u|^{p-2}\nabla u \quad \text{in } L^{p/p-1}(\Omega) \end{array}\right\} \qquad (1.4.11)$$

   at least for a subsequence.

1.4 A Survey on p-Harmonic Maps

Outline of the proof: We assume $u_n \rightharpoonup u$ in $H^{1,p}(\Omega, \mathbb{R}^k)$ and in addition that (1.4.10) holds. It is then sufficient to show

$$\nabla u_n(x) \to \nabla u(x) \quad \text{a.e.} \tag{1.4.12}$$

for a subsequence. In fact the second statement in (1.4.11) and hence also the claim of Theorem 1.4.8 are direct consequences of (1.4.12) and Egoroff's theorem. For (1.4.12) we have to prove:

Lemma 1.4.9 *Consider an arbitrary subregion $\Omega^* \subset\subset \Omega$. Then*

$$\lim_{n \to \infty} \mathcal{L}^m(\{x \in \Omega^* : |\nabla u_n(x) - \nabla u(x)| \geq \alpha\}) = 0$$

is valid for any $\alpha > 0$.

To this purpose we choose a sequence $\{\Omega_\ell\}$ of measurable sets and a suitable subsequence of $\{u_n\}$ with the following properties:

(i) $\quad u_n \to u$ strongly in $L^p(\Omega)$ and a.e.,

(ii) $\quad \Omega_\ell \subset \Omega_{\ell+1}$, $\mathcal{L}^m\left(\Omega - \bigcup_{\ell=1}^\infty \Omega_\ell\right) = 0$,

(iii) $\quad \lim_{n \to \infty} \|u_n - u\|_{L^\infty(\Omega_\ell)} = 0$, especially $\|u_n - u\|_{L^\infty(\Omega_\ell)} \leq 1/\ell$ for $\ell \geq n_\ell$.

If Ω^* is given we pick $\eta \in C_0^1(\Omega, [0,1])$ with $\eta = 1$ on Ω^* and claim

$$\limsup_{n \to \infty} \int_{\Omega_\ell} \eta(|\nabla u_n|^{p-2}\nabla u_n - |\nabla u|^{p-2}\nabla u) \cdot \nabla(u_n - u)\, dx \xrightarrow[\ell \to \infty]{} 0. \tag{1.4.13}$$

It is an easy matter to check that (1.4.13) implies Lemma 1.4.9. On the other hand weak convergence $u_n \rightharpoonup u$ in $H^{1,p}(\Omega, \mathbb{R}^k)$ and the choice of Ω_ℓ give

$$0 \leq \limsup_{n \to \infty} \int_{\Omega_\ell} \ldots\, dx$$

$$= \limsup_{n \to \infty} \int_{\Omega_\ell} |\nabla u_n|^{p-2}\nabla u_n \cdot \nabla(u_n - u)\, \eta\, dx$$

$$= \limsup_{n \to \infty} \int_{\Omega_\ell} |\nabla u_n|^{p-2}\nabla u_n \cdot \nabla[u_n - u]^L \eta\, dx =: a_\ell$$

with $L := \frac{1}{\ell}$ and $v^L := \begin{cases} v & \text{if } |v| \leq L \\ \frac{L}{|v|}v & \text{if } |v| \geq L \end{cases}$.

Finally we use assumption (1.4.10) in order to show $\lim_{\ell \to \infty} a_\ell = 0$. For further details we again refer to [38].

1.5 Variational Inequalities and Asymptotically Regular Integrands

In sections 1.2 and 1.3 we considered the obstacle problem $\int_\Omega |\nabla u|^p dx \to \text{Min}$ subject to a constraint of the form $u(x) \in \overline{M}$ for a fixed region $M \subset \mathbb{R}^N$. The results obtained there do not cover the case of side conditions of the form $u^i \geq \psi^i$, $i = 1, \ldots, N$, with given functions ψ^i. The following statements are taken from [33].

Theorem 1.5.1 *Suppose that $\psi \in H^{1,p}(\Omega, \mathbb{R}^N)$, Ω a domain in $\mathbb{R}^n$, satisfies a Morrey condition of the form*

$$\int_{B_R(x)} |\nabla \psi|^p \, dx \leq K \cdot R^{n-p+p\cdot\alpha} \tag{1.5.1}$$

for all balls $B_R(x) \subset \Omega$ with $K > 0$ and $\alpha \in (0,1)$. Then, if $u \in H^{1,p}(\Omega, \mathbb{R}^N)$ is a local minimizer of $\int_\Omega |\nabla u|^p \, dx$ subject to $u^i(x) \geq \psi^i(x)$ a.e., $i = 1, \ldots, N$, u is of class $C^{0,\alpha}(\Omega, \mathbb{R}^N)$.

Remark: (1.5.1) implies $\psi \in C^{0,\alpha}(\Omega, \mathbb{R}^N)$. Clearly it would be sufficient to replace (1.5.1) by a local condition saying that (1.5.1) holds for balls $B_R(x)$ contained in a subregion Ω^* of Ω with $K = K(\Omega^*)$ depending on Ω^*.

Sketch of the proof: Fix a ball $B_R \subset \Omega$ and let $w \in H^{1,p}(B_R, \mathbb{R}^N)$ denote the unique minimizer of p–energy $\int_{B_R} |\nabla w|^p \, dx$ in the class $(u - \psi) + \overset{\circ}{H}{}^{1,p}(B_R, \mathbb{R}^N)$. We have

$$w^i = u^i - \psi^i \geq 0 \quad \text{on } \partial B_R$$

and one easily checks the maximum principle

$$w^i \geq 0 \quad \text{on } B_R, \, i = 1, \ldots, N.$$

Hence $v := \begin{cases} u & \text{on } \Omega - B_R \\ w + \psi & \text{on } B_R \end{cases}$ is an admissible comparison function, i.e. $\int_{B_R} |\nabla u|^p \, dx \leq \int_{B_r} |\nabla (u + t(v - u))|^p \, dx$ for any $0 \leq t \leq 1$ so that

$$\int_{B_R} |\nabla u|^{p-2} \nabla u \cdot \nabla (v - u) \, dx \geq 0. \tag{1.5.2}$$

For any ball we obtain $(r < R)$

1.5 Variational Inequalities and Asymptotically Regular Integrands

$$\int_{B_r} |\nabla u|^p \, dx \leq c_1 \left\{ \int_{B_r} |\nabla u - \nabla v|^p \, dx + \int_{B_r} |\nabla v|^p \, dx \right\}$$

$$\leq c_2 \left\{ \int_{B_R} |\nabla u - \nabla v|^p \, dx + \int_{B_r} |\nabla w|^p \, dx + \int_{B_r} |\nabla \psi|^p \, dx \right\}$$

$$\leq c_3 \cdot \left\{ \int_{B_R} |\nabla u - \nabla v|^p \, dx + \left(\frac{r}{R}\right)^n \int_{B_R} |\nabla w|^p \, dx \right.$$

$$\left. + \int_{B_R} |\nabla \psi|^p \, dx \right\}$$

having used the Uhlenbeck estimate for ∇w. Next observe

$$\int_{B_R} |\nabla w|^p \, dx \leq \int_{B_R} |\nabla (u - \psi)|^p \, dx$$

$$\leq c_4 \left\{ \int_{B_R} |\nabla u|^p \, dx + \int_{B_R} |\nabla \psi|^p \, dx \right\}$$

and

$$\int_{B_R} |\nabla u - \nabla v|^p \, dx$$

$$\leq c_5 \cdot \left\{ \int_{B_R} |\nabla u|^{p-2} \nabla u \cdot \nabla (u - v) \, dx \right.$$

$$\left. - \int_{B_R} |\nabla v|^{p-2} \nabla v \cdot \nabla (u - v) \, dx \right\}$$

$$\underset{(1.5.1)}{\leq} - c_5 \cdot \int_{B_R} |\nabla v|^{p-2} \nabla v \cdot \nabla (u - v) \, dx$$

$$= c_5 \int_{B_R} \left(|\nabla v|^{p-2} \nabla v - |\nabla w|^{p-2} \nabla w \right) \cdot \nabla (u - v) \, dx$$

$$\leq c_6 \int_{B_R} |\nabla v - \nabla w| \cdot \left\{ |\nabla v|^{p-2} + |\nabla w|^{p-2} \right\} \cdot |\nabla u - \nabla v| \, dx$$

$$\leq c_7 \int_{B_R} \left(|\nabla \psi|^{p-1} + |\nabla \psi| \cdot |\nabla w|^{p-2} \right) \cdot |\nabla u - \nabla v| \, dx$$

($|\nabla v| \leq |\nabla w| + |\nabla \psi|$ by definition of v).
With Young's inequality we get

$$\int_{B_R} |\nabla u - \nabla v|^p \, dx \leq c_8 \left(\delta \cdot \int_{B_R} |\nabla u - \nabla v|^p \, dx + c(\delta) \left\{ \int_{B_R} |\nabla \psi|^p \, dx \right. \right.$$

$$\left. \left. + \int_{B_R} \left(|\nabla \psi| \cdot |\nabla w|^{p-2} \right)^{\frac{p}{p-1}} \, dx \right\} \right)$$

so that after appropriate choice of δ

$$\int_{B_R} |\nabla u - \nabla v|^p \, dx \leq c_9 \cdot \left\{ \int_{B_R} |\nabla \psi|^p \, dx + \int_{B_R} \left(|\nabla \psi| \cdot |\nabla w|^{p-2} \right)^{\frac{p}{p-1}} dx \right\}$$

$$\leq c_{10} \left\{ \varepsilon \cdot \int_{B_R} |\nabla w|^p \, dx + c(\varepsilon) \int_{B_R} |\nabla \psi|^p \, dx \right\}$$

$$\leq c_{11} \left\{ \varepsilon \cdot \int_{B_R} |\nabla u|^p \, dx + c(\varepsilon) \int_{B_R} |\nabla \psi|^p \, dx \right\}$$

for any $\varepsilon > 0$. Here $c(\varepsilon) \nearrow \infty$ as $\varepsilon \searrow 0$. Inserting all the different estimates and using (1.5.1) we end up with

$$\int_{B_r} |\nabla u|^p \, dx \leq c_{12} \left\{ \left(\varepsilon + \left(\frac{r}{R} \right)^n \right) \int_{B_R} |\nabla u|^p \, dx + c(\varepsilon) \int_{B_R} |\nabla \psi|^p \, dx \right\}$$

$$\leq c_{13} \left\{ \left(\varepsilon + \left(\frac{r}{R} \right)^n \right) \int_{B_R} |\nabla u|^p \, dx + c(\varepsilon) R^{n-p+p \cdot \alpha} \right\}$$

for any $r < R$ and $\varepsilon > 0$. This completes the proof of the Theorem. $\square$

Next we prove global $C^1(\Omega)$–regularity for local minimizers in the scalar case.

Theorem 1.5.2 *Suppose $N = 1$ and $\psi \in H^{2,\infty}_{\text{loc}}(\Omega)$. Let $u \in H^{1,p}_{\text{loc}}(\Omega)$ denote a local minimizer of $\int_{\Omega} |\nabla u|^p \, dx$ subject to the constraint $u \geq \psi$ a.e. Then $u \in C^{1,\mu}(\Omega)$ for some $0 < \mu < 1$.*

Remarks:

1. As we shall see below our proof works only in the scalar case $N = 1$ but I think by working directly with the local minimality of u instead of using the Euler equation below it should be possible to extend Theorem 1.5.2 to several constraints $u^i \geq \psi^i$, $i = 1, \ldots, N$. One should discuss the details.

2. But even in case $N = 1$ Theorem 1.5.2 was known only in the dimension $n = 2$. In this case a proof is due to [65] using quasiconformal mappings.

Lemma 1.5.1 *Under the assumptions of Theorem 1.5.2 the following equation holds ($\eta \in C_0^1(\Omega)$)*

$$\int_{\Omega} |\nabla u|^{p-2} \nabla u \cdot \nabla \eta \, dx = \int_{\Omega} \Theta \eta \cdot \left(-\partial_{\alpha} (|\nabla \psi|^{p-2} \partial_{\alpha} \psi) \right)^+ dx$$

where $(\ldots)^+$ is the positive part and $\Theta : \Omega \to [0,1]$ is $\mathcal{L}^n$–measurable.

Remarks:

1. Using the continuity of u as well as the information

$$|\nabla u|^{p-2} \nabla u \in H^{1,\frac{p}{p-1}}_{loc}(\Omega)$$

   (which follows by using difference quotient techniques) one can prove the equation

   $$* \qquad \partial_\alpha(|\nabla u|^{p-2} \partial_\alpha u) = \mathbf{1}_{[u=\psi]} \partial_\alpha(|\nabla \psi|^{p-2}\partial_\alpha \psi)$$

   and

   $$\partial_\alpha(|\nabla \psi|^{p-2}\partial_\alpha \psi) \leq 0 \quad \text{a.e. on } [u=\psi].$$

2. In the vectorial case $*$ reads

   $$\partial_\alpha(|\nabla u|^{p-2} \partial_\alpha u^i) = \mathbf{1}_{[u^i=\psi^i]} \partial_\alpha(|\nabla u|^{p-2} \partial_\alpha \psi^i), \quad i=1,\ldots,N.$$

3. Note that our proof of Lemma 1.5.1 does not use $u \in C^0(\Omega)$.

Proof of Lemma 1.5.1: (compare Theorem 1.2.1) Let $u_t := u+t\cdot\eta\cdot h_\varepsilon(u-\psi)$, $\eta \in C^1_0(\Omega)$, $\eta \geq 0$, $t > 0$, $\varepsilon > 0$, $0 \leq h_\varepsilon \leq 1$, $h_\varepsilon = 1$ on $(0,\varepsilon)$, $h_\varepsilon \equiv 0$ on $(2\varepsilon, \infty)$, $h'_\varepsilon \leq 0$. Since

$$\frac{1}{t}\left(\int_\Omega |\nabla u_t|^p\, dx - \int_\Omega |\nabla u|^p\, dx\right) \geq 0$$

we deduce by passing to the limit $t \searrow 0$

$$0 \leq \int_\Omega |\nabla u|^{p-2}\nabla u \cdot \nabla(\eta \cdot h_\varepsilon(u-\psi))\, dx$$

and there is a Radon measure $\lambda \geq 0$ on Ω such that

$$\int_\Omega |\nabla u|^{p-2}\nabla u \nabla(\eta \cdot h_\varepsilon(u-\psi))\, dx = \int_\Omega \eta\, d\lambda$$

now for all $\eta \in C^1_0(\Omega)$. Clearly λ is independent of ε and in order to get a bound on λ we fix $\eta \geq 0$ and pass to the limit $\varepsilon \searrow 0$.

$$\int_\Omega \eta\, d\lambda = \int_\Omega |\nabla u|^{p-2}\nabla u \cdot \nabla\eta \cdot h_\varepsilon(u-\psi)\, dx$$
$$+ \int_\Omega |\nabla u|^{p-2} h'_\varepsilon(u-\psi) \cdot \nabla u \cdot \nabla(u-\psi)\, \eta\, dx$$
$$= \int_\Omega |\nabla u|^{p-2}\nabla u \cdot \nabla\eta \cdot h_\varepsilon(u-\psi)\, dx$$
$$+ \int_\Omega (|\nabla u|^{p-2}\nabla u - |\nabla\psi|^{p-2}\nabla\psi)\, \eta \cdot h'_\varepsilon(u-\psi) \cdot (\nabla u - \nabla\psi)\, dx$$
$$+ \int_\Omega |\nabla\psi|^{p-2}\nabla\psi \cdot \eta \cdot h'_\varepsilon(u-\psi) \cdot \nabla(u-\psi)\, dx$$
$$\leq \int_\Omega |\nabla u|^{p-2}\nabla u \cdot \nabla\eta \cdot h_\varepsilon(u-\psi)\, dx$$
$$+ \int_\Omega |\nabla\psi|^{p-2}\nabla\psi \cdot \nabla(u-\psi)\, \eta \cdot h'_\varepsilon(u-\psi)\, dx$$
$$= \int_\Omega |\nabla u|^{p-2}\nabla u \cdot \nabla\eta\, h_\varepsilon(u-\psi)\, dx$$
$$+ \int_\Omega |\nabla\psi|^{p-2}\nabla\psi \cdot \nabla(h_\varepsilon(u-\psi))\, \eta\, dx$$
$$= \int_\Omega \ldots dx + \int_\Omega |\nabla\psi|^{p-2}\nabla\psi \cdot \nabla(\eta \cdot h_\varepsilon(u-\psi))\, dx$$
$$- \int_\Omega |\nabla\psi|^{p-2}\nabla\psi \cdot \nabla\eta \cdot h_\varepsilon(u-\psi)\, dx$$
$$= \int_\Omega \left[|\nabla u|^{p-2}\nabla u - |\nabla\psi|^{p-2}\nabla\psi\right] \cdot \nabla\eta \cdot h_\varepsilon(u-\psi)\, dx$$
$$- \int_\Omega \partial_\alpha(|\nabla\psi|^{p-2}\partial_\alpha\psi) \cdot \eta \cdot h_\varepsilon(u-\psi)\, dx$$
$$\xrightarrow{\varepsilon\downarrow 0} \int_{[u=\psi]} \left[|\nabla u|^{p-2}\nabla u - |\nabla\psi|^{p-2}\nabla\psi\right] \cdot \nabla\eta\, dx$$
$$+ \int_{[u=\psi]} (-\partial_\alpha(|\nabla\psi|^{p-2}\partial_\alpha\psi)) \cdot \eta\, dx\,.$$

Since $\nabla u = \nabla\psi$ a.e. on $[u=\psi]$ we arrive at

$$\int_\Omega \eta\, d\lambda \leq \int_{[u=\psi]} (-\partial_\alpha(|\nabla\psi|^{p-2}\partial_\alpha\psi))\, \eta\, dx\,,$$

especially $\mathbf{1}_{[u=\psi]} \cdot (-\partial_\alpha(|\nabla\psi|^{p-2}\partial_\alpha\psi)) \geq 0$ a.e. and λ takes the form

1.5 Variational Inequalities and Asymptotically Regular Integrands

$$\lambda = \mathcal{L}^n \llcorner \{\Theta \cdot (-\partial_\alpha(|\nabla\psi|^{p-2}\partial_\alpha\psi))^+\}$$

for a density function Θ. □

The proof of Theorem 1.5.2 now is immediate: we have

$$\partial_\alpha(|\nabla u|^{p-2}\partial_\alpha u) = f$$

on Ω for some function $f \in L^\infty_{\text{loc}}(\Omega)$ so that the local comparison technique used in Theorem C applies with obvious simplifications. It should also be noted that the above equation provides us with an alternative proof of Theorem 1.5.1 at least for $N = 1$. □

We finish our discussion of scalar variational inequalities with a result concerning the structure of the contact set $[u = \psi]$.

Theorem 1.5.3 *Suppose that ψ is of class $C^{3,\beta}(\Omega)$ for some $\beta > 0$ with $\nabla\psi(x) \neq 0$ everywhere and in addition strictly concave in the sense that*

$$\frac{\partial^2\psi}{\partial x_\alpha \partial x_\beta}(x)\eta_\alpha\eta_\beta \leq -\mu(x)|\eta|^2$$

for $x \in \Omega$, $\eta \in \mathbb{R}^n$ with $\mu(x) > 0$. (Actually it would be sufficient to have $\partial_\alpha(|\nabla\psi|^{p-2}\partial_\alpha\psi) < 0$.) Let $u \in H^{1,p}(\Omega)$ denote a local minimizer of $\int_\Omega |\nabla u|^p dx$ under the side condition $u \geq \psi$ a.e. Then the contact set $I := \{x \in \Omega : u(x) = \psi(x)\}$ is of locally finite perimeter in Ω, i.e. $\mathbf{1}_I \in BV_{\text{loc}}(\Omega)$.

Remarks:

1. The proof is rather involved and uses a penalty method, for details we refer to [33].

2. For the vectorial case we have a weaker result which is given in the paper [28]. □

Next we draw our attention to *systems of variational inequalities with natural growth* (compare [36], [37]). In order to simplify things we look at the following model problem:

$$\begin{cases} \text{find } u \in \mathbb{K} \text{ such that} \\ \int_\Omega |\nabla u|^{p-2}\nabla u \cdot \nabla(v-u)dx \geq \int_\Omega f(\cdot,u,\nabla u)\cdot(v-u)dx \\ \text{holds for all } v \in \mathbb{K}. \end{cases} \quad (1.5.3)$$

Here $\Omega \subset \mathbb{R}^n$, $n \geq 2$, is a bounded domain, K denotes a compact *convex* region $\subset \mathbb{R}^N$ which is the closure of a C^2–domain and the convex class $\mathbb{K}$ is defined by

$$\mathbb{K} = \{u \in H^{1,p}(\Omega, \mathbb{R}^N) : u(x) \in K \text{ a.e.}, u = u_0 \text{ on } \partial\Omega\}$$

for a given function $u_0 \in H^{1,p}(\Omega, \mathbb{R}^N)$, $u_0(x) \in K$. f is a continuous function on $\Omega \times \mathbb{R}^N \times \mathbb{R}^{nN}$ such that

$$|f(x, y, A)| \leq a \cdot |A|^p \qquad (1.5.4)$$

for a positive constant a.

Theorem 1.5.4 *In addition to (1.5.4) assume*

$$a < 1/\operatorname{diam}(K). \qquad (1.5.5)$$

Then (1.5.3) has at least one solution $u \in \mathbb{K}$ which is of class $C^{1,\varepsilon}(\Omega - \Sigma)$ for some relatively closed subset Σ of Ω with $\mathcal{H}^{n-p}(\Sigma) = 0$.

Comments:

1. In case $p = 2$ this result is due to Hildebrandt and Widman [62]. They proved $\Sigma = \emptyset$ and apriori Hölder estimates for solutions of (1.5.3) which in turn can be used to obtain existence. Unfortunately they use Green's function technique which is limited to the case $p = 2$.

2. Similar to Theorem 1.4.2 we conjecture $\operatorname{Sing} u = \emptyset$ for any solution u of (1.5.3) under condition (1.5.5) but apart from some rather artificial cases we were unable to exclude singular points.

3. It should be noted that (1.5.3) does not have variational structure in the sense that we require (1.5.3) to be associated to some convex minimisation problem "$\mathcal{F}(u) \to \text{Min in } \mathbb{K}$". In this case the existence of solutions would be trivial whereas regularity would be a consequence of our Theorems contained in section 1.1.

The proof of Theorem 1.5.4 proceeds in several steps

(1) *Approximation:* For $k \in \mathbb{N}$ we define

$$f_k(x, y, Q) := \begin{cases} f(x, y, Q), & |f(x, y, Q)| \leq k \\ k \cdot f(x, y, Q)/|f(x, y, Q)|, & \text{otherwise} \end{cases}$$

and look at the problem

1.5 Variational Inequalities and Asymptotically Regular Integrands 65

$$\begin{cases} \text{find } \tilde{u} \in \mathbb{K} \text{ such that} \\ \int_\Omega |\nabla \tilde{u}|^{p-2} \nabla \tilde{u} \cdot \nabla(v - \tilde{u}) dx \geq \int_\Omega f_k(\cdot, \tilde{u}, \nabla \tilde{u}) \cdot (v - \tilde{u}) dx \\ \text{holds for all } v \in \mathbb{K}. \end{cases} \quad (1.5.3)_k$$

In order to obtain a solution $u_k = \tilde{u}$ of $(1.5.3)_k$ let

$$\begin{cases} T: \mathbb{K} \to \mathbb{K}, \\ w := T(u) \text{ is the unique solution of} \\ \int_\Omega \left(|\nabla w|^{p-2} \nabla w \cdot (\nabla \varphi - \nabla w) - f_k(\cdot, u, \nabla u) \cdot (\varphi - w) \right) dx \geq 0 \\ \text{for all } \varphi \in \mathbb{K}. \end{cases}$$

Clearly T is well defined and $T(\mathbb{K})$ is precompact, i.e.: for any sequence $\{u_i\} \subset \mathbb{K}$, $w_i := Tu_i$ has a subsequence converging strongly in $H^{1,p}(\Omega, \mathbb{R}^N)$. Since $u_0 \in \mathbb{K}$ we first deduce

$$\int_\Omega |\nabla w_i|^{p-2} \nabla w_i \cdot (\nabla u_0 - \nabla w_i) dx \geq \int_\Omega f_k(\cdot, u_i, \nabla u_i) \cdot (u_0 - w_i) dx$$

so that

$$\int_\Omega |\nabla w_i|^p dx \leq \int_\Omega |\nabla w_i|^{p-1} \cdot |\nabla u_0| dx + k \cdot \mathcal{L}^n(\Omega) \cdot 2 \cdot \operatorname{diam}(K)$$

and in conclusion $\sup_i \|w_i\|_{H^{1,p}(\Omega)} < \infty$. Thus we have for a subsequence

$$\begin{aligned} w_i &\rightharpoonup w && \text{weakly in } H^{1,p}(\Omega, \mathbb{R}^N) \\ w_i &\to w && \text{strongly in } L^p(\Omega, \mathbb{R}^N). \end{aligned}$$

On the other hand by definition of T

$$\int_\Omega \left(|\nabla w_i|^{p-2} \nabla w_i \cdot \nabla(w_j - w_i) - f_k(\cdot, u_i, \nabla u_i)(w_j - w_i) \right) dx \geq 0,$$

$$\int_\Omega \left(|\nabla w_j|^{p-2} \nabla w_j \cdot \nabla(w_i - w_j) - f_k(\cdot, u_j, \nabla u_j)(w_i - w_j) \right) dx \geq 0$$

which implies

$$\int_\Omega \left(|\nabla w_i|^{p-2} \nabla w_i - |\nabla w_j|^{p-2} \nabla w_j|^{p-2} \nabla w_j \right) \cdot (\nabla w_i - \nabla w_j) dx$$

$$\leq \int_\Omega f_k(\cdot, u_i, \nabla u_i) \cdot (w_i - w_j) dx - \int_\Omega f_k(\cdot, u_j, \nabla u_j) \cdot (w_i - w_j) dx,$$

therefore

$$\int_\Omega |\nabla w_i - \nabla w_j|^p \, dx \le c(n,N,p) \cdot k \cdot \int_\Omega |w_i - w_j| \, dx \xrightarrow[i,j\to\infty]{} 0$$

so that $\{w_i\}$ is a Cauchy sequence in $H^{1,p}(\Omega, \mathbb{R}^N)$. According to Schauder's fixed point theorem we find $\tilde u \in \mathbb{K}$ s.t. $\tilde u = T\tilde u$. $\tilde u =: u_k$ then is a solution of the approximate problem $(1.5.3)_k$. After passing to a subsequence (use $(1.5.3)_k$ with $v = u_0$ and recall $(1.5.5)$) we may assume $u_k \rightharpoonup: u$ in $H^{1,p}(\Omega, \mathbb{R}^N)$ and $u_k \to u$ in $L^p(\Omega, \mathbb{R}^N)$ for some $u \in \mathbb{K}$.

(2) *Uniform partial regularity:*

Proposition: *There are constants $\varepsilon > 0$, $\alpha \in (0,1)$ and $C > 0$ independent of k as follows: if for some ball $B_R(x) \subset \Omega$ we have*

$$\fint_{B_R(x)} |u - (u)_R|^p \, dx < \varepsilon$$

then

$$u, u_k \in C^{1,\alpha}(B_{R/2}(x), \mathbb{R}^N)$$

and

$$|\nabla u_k(x_1) - \nabla u_k(x_2)| + |\nabla u(x_1) - \nabla u(x_2)| \le C \cdot |x_1 - x_2|^\alpha$$

for all $x_1, x_2 \in B_{R/2}(x)$. Moreover

$$u_k \to u \quad \text{in } C^{1,\alpha}(B_{R/2}(x)).$$

Proof: Suppose that w is a solution of

$$\int_\Omega |\nabla w|^{p-2} \nabla w \cdot \nabla(v-w) \, dx \ge \int_\Omega g(\cdot, w, \nabla w) \cdot (v-w) \, dx$$

for all $v \in \mathbb{K}$ where g satisfies

$$|g(x,y,Q)| \le a \cdot |Q|^p$$

with

$$a < 1/\operatorname{diam} K.$$

This holds for $w := u_k$, $g := f_k$. (Up to now we do not know that the limit function u is a solution of $(1.5.3)$ since we only have weak convergence.)

Since the side condition is convex and by using the smallness condition $(1.5.5)$ we immediately deduce Caccioppoli's inequality and by the way $\nabla w \in L^t_{\text{loc}}(\Omega)$ for some absolute $t > p$ (independent of w). If $B_R(x)$ is any ball in Ω and if $h \in H^{1,p}(B_R(x), \mathbb{R}^N)$ is the minimizer of $\int_{B_R(x)} |\nabla h|^p \, dx$ for boundary values w then

1.5 Variational Inequalities and Asymptotically Regular Integrands

$$\tilde{h} := \begin{cases} w & \text{on } \Omega - B_R(x) \\ h & \text{on } B_R(x) \end{cases}$$

is in $\mathbb{K}$ and we may proceed as in section 1.2 to show

$$\begin{cases} R^{p-n} \int_{B_R(x)} |\nabla w|^p \, dx < \varepsilon_1 \Rightarrow \\ \nabla w \in C^{0,\alpha}(B_{R/2}(x)) \text{ uniformly} \end{cases} \quad (1.5.6)$$

where ε_1 is independent of w. By Caccioppoli's inequality we see that the conclusion of (1.5.6) holds under the assumption

$$\fint_{B_R(x)} |w - (w)_R|^p \, dx < \varepsilon. \quad (1.5.7)$$

Now choose a ball $B_R(x) \subset \Omega$ such that (1.5.7) holds for the limit function. For $k \gg 1$ (1.5.7) is valid for u_k and we deduce $\nabla u_k \in C^{0,\alpha}(B_{R/2}(x))$ uniformly. The statement of the Proposition now follows from Arzela's Theorem. □

(3) *Solution of the variational inequality (1.5.3):* Let

$$\Sigma := \{x \in \Omega : \liminf_{r \downarrow 0} \fint_{B_r(x)} |u - (u)_r|^p \, dz > 0\}.$$

From the Proposition we deduce $u_k \to u$ in $C^{1,\alpha}$ on compact subsets of $\Omega - \Sigma$.

Consider a ball $B_R(x_0) \subset\subset \Omega - \Sigma$ and a function $w \in \mathbb{K}$ such that $\mathrm{spt}\,(u - w) \subset\subset B_R(x_0)$. For $\eta \in C_0^1(B_R(x_0), [0,1])$, $\eta \equiv 1$ on $B_{R-\delta}(x_0)$ we apply $(1.5.3)_k$ to
$v := (1-\eta)u_k + \eta w$ and get after passing to the limit $k \to \infty$

$$\int \left(|\nabla u|^{p-2} \nabla u \cdot \nabla(\eta[w-u]) - f(\cdot, u, \nabla u) \cdot (w-u)\eta \right) dx \geq 0$$

Since $\mathrm{spt}\,(u - w)$ is compact in $B_R(x_0)$ we can arrange $\eta = 1$ on $\mathrm{spt}\,(u - w)$ so that

$$\int_{B_R(x_0)} \left(|\nabla u|^{p-2} \nabla u \cdot \nabla(w-u) - f(\cdot, u, \nabla u) \cdot (w-u) \right) dx \geq 0 \quad (1.5.8)$$

$$\forall w \in \mathbb{K}, \quad \mathrm{spt}\,(w - u) \subset\subset B_R(x_0).$$

Up to now we have not made use of the smoothness of ∂K: as in Theorem 1.2.1 (use $w := u + \eta \cdot h_\varepsilon(d(u)) \mathcal{N}(u)$ as normal variation and an appropriate tangential one, compare Theorem 1.2.1 for notations) we deduce from (1.5.8)

$$\begin{cases} \int_{B_R(x_0)} \left(|\nabla u|^{p-2} \nabla u \cdot \nabla \psi - f(\cdot, u, \nabla u) \cdot \psi \right) dx \\ = \int_{B_R(x_0) \cap [u \in \partial K]} \psi \cdot \mathcal{N}(u) \, B(\cdot, u, \nabla u) dx \end{cases} \quad (1.5.9)$$

for all $\psi \in C_0^1(B_R(x_0), \mathbb{R}^N)$ where $\mathcal{N}(y)$ is the interior normal to ∂K and $B(\cdot, u, \nabla u) \geq 0$ a.e. on $[u \in \partial K]$ with the property $B(\cdot, u, \nabla u) \leq \tilde{a} \cdot |\nabla u|^p$ for some absolute constant $\tilde{a}$. By a covering argument (1.5.9) extends to all $\psi \in C_0^1(\Omega - \Sigma, \mathbb{R}^N)$. Next we want to show (1.5.6) for all $\psi \in C_0^1(\Omega, \mathbb{R}^N)$. To this purpose observe $\mathcal{H}^{n-p}(\Sigma) = 0$, hence $\text{cap}_p(\Sigma) = 0$ and we find a sequence $\eta_\nu \in C^\infty(\mathbb{R}^n, [0,1])$ such that $\text{spt}\, \eta_\nu \cap \Sigma = \emptyset$, $\eta_\nu \xrightarrow[\nu \to \infty]{} 1$ a.e. and $\int_{\mathbb{R}^n} |\nabla \eta_\nu|^p \, dx \xrightarrow[\nu \to \infty]{} 0$. (Compare [73].) Then, if $\psi \in C_0^1(\Omega, \mathbb{R}^N)$ is arbitrary, we may apply (1.5.9) to $\eta_\nu \cdot \psi$ which finally implies

$$\begin{cases} \int_\Omega \left(|\nabla u|^{p-2} \nabla u \cdot \nabla \psi - f(\cdot, u, \nabla u) \cdot \psi \right) dx \\ = \int_{\Omega \cap [u \in \partial K]} \psi \cdot \mathcal{N}(u) \, B(\cdot, u, \nabla u) \, dx \end{cases} \quad (1.5.10)$$

for all $\psi \in C_0^1(\Omega, \mathbb{R}^N)$ and by a density argument for $\psi \in \overset{\circ}{H}{}^{1,p} \cap L^\infty(\Omega, \mathbb{R}^N)$. We apply (1.5.10) to $\psi := v - u$ with $v \in \mathbb{K}$. Since K is convex we have $(v - u) \cdot \mathcal{N}(u) \geq 0$ a.e. on $[u \in \partial K]$, thus (1.5.10) turns into the desired variational inequality (1.5.3). $\square$

During these lectures we have been confronted several times with degenerate systems

$$- \partial_\alpha (|\nabla u|^{p-2} \partial_\alpha u) = f(\cdot, u, \nabla u) \quad (1.5.11)$$

on a domain $\Omega \subset \mathbb{R}^n$ with f of natural growth, i.e.

$$|f(x, y, Q)| \leq a \cdot |Q|^p \quad (1.5.12)$$

where for technical simplicity we drop the lower order terms. As an application of Theorem 1.5.4 we obtain an existence theorem for the Dirichlet problem associated to (1.5.11).

Theorem 1.5.5 *Suppose that $u_0 \in H^{1,p}(\Omega, \mathbb{R}^N) \cap L^\infty$ is given satisfying*

$$a < \frac{1}{2 \cdot \|u_0\|_\infty} \, . \quad (1.5.13)$$

Then there exists a solution $u \in H^{1,p}(\Omega, \mathbb{R}^N) \cap L^\infty$ of (1.5.11) for boundary values u_0 which is of class $C^{1,\alpha}(\Omega - \Sigma)$ up to a relatively closed singular set with $\mathcal{H}^{n-p}(\Sigma) = 0$.

1.5 Variational Inequalities and Asymptotically Regular Integrands

Remarks: For $p = 2$ it is possible to show Sing $u = \emptyset$ even under the weaker condition $a < \frac{1}{\|u_0\|_\infty}$.

Proof of Theorem 1.5.5: For $R := \|u_0\|_\infty$ define $K = \{z \in \mathbb{R}^N : |z| \leq R + \varepsilon\}$ where $\varepsilon > 0$ is chosen (compare (1.5.13)) in such a way that $a < \frac{1}{\operatorname{diam} K}$ holds. According to Theorem 1.5.4 we find $u \in H^{1,p}(\Omega, K)$ such that $u = u_0$ on $\partial \Omega$ and

$$\int_\Omega |\nabla u|^{p-2} \nabla u \cdot \nabla (v - u) dx \geq \int_\Omega f(\cdot, u, \nabla u) \cdot (v - u) dx$$

for all $v \in H^{1,p}(\Omega, K)$, $v = u_0$ on $\partial \Omega$, and u has the stated regularity properties.

For $\eta \in C_0^1(\Omega, [0, 1])$ we let $v := (1 - \eta) \cdot u$ and get after a simple calculation

$$\int_\Omega |\nabla u|^{p-2} \nabla |u|^2 \cdot \nabla \eta \, dx \leq 0 \, .$$

By approximation this inequality extends to $\eta \in \mathring{H}^{1,p}(\Omega)$, $\eta \geq 0$, especially we may insert $\eta := \max(0, |u|^2 - R^2)$ which gives $\int_{[|u|>R]} |\nabla u|^{p-2} |\nabla \eta|^2 \, dx = 0$ so that $\|u\|_\infty \leq R$. $u + t \cdot \psi$, $|t| << 1$, $\psi \in C_0^1(\Omega, \mathbb{R}^N)$, then belongs to the admissible class so that u is a solution of (1.5.11). $\square$

We close this section with some recent regularity results taken from [39] concerning minimizers of variational integrals

$$\mathcal{F}(u, \Omega) := \int_\Omega f(\nabla u) \, dx$$

with integrand $f : \mathbb{R}^{nN} \to \mathbb{R}$ satisfying the asymptotic condition

$$\lim_{|A| \to \infty} |A|^{-p} \cdot f(A) \quad \text{exists in } (0, \infty) \tag{1.5.14}$$

for some power $p \in (1, \infty)$. We require neither convexity nor differentiability of f and prove

Theorem 1.5.6 *Let f denote a continuous function with (1.5.14) and suppose that $u \in H^{1,p}(\Omega, \mathbb{R}^N)$ locally minimizes $\mathcal{F}(\cdot, \Omega)$. Then $u \in C^{0,\mu}(\Omega, \mathbb{R}^N)$ for any $0 < \mu < 1$. Moreover, u is differentiable at almost all points of Ω.*

Remarks:

1. In the physical case $n = N = 3$ we can treat integrands f of the form
$$\begin{aligned}f(A) :=\ & |A|^p + c_1|A|^q + c_2 \cdot |\operatorname{Adj} A|^\alpha \\ & + c_3 \cdot |\det A|^\beta + c_4 \cdot (\delta + |\det Q|)^{-s}\end{aligned}$$
with $q < p$, $0 < \alpha < \frac{p}{2}$, $0 < \beta < \frac{p}{3}$, $\delta, s > 0$ where the last term is included to indicate that "deformations" u with small determinant give large contributions to the energy. This is a wellknown phenomenon in nonlinear elasticity.

2. We do not claim that the set of points $x \in \Omega$, where $\nabla u(x)$ exists, is open. It is only known that the complement of these points is of vanishing $\mathcal{L}^n$–measure. We conjecture that u is Lipschitz on Ω but we were not able to prove this fact.

3. Since f is not assumed to be convex the existence of $\mathcal{F}$–minimizers is not guaranteed. But the spirit of Theorem 1.5.6 lies in applications to *problems of relaxation:*
Let f satisfy (1.5.14) and define (quasiconvex envelope)
$$Qf := \sup\{\tilde{f} : \tilde{f} \leq f,\, \tilde{f} \text{ quasiconvex}\}\,.$$
Then (1.5.14) holds for Qf, moreover Qf is quasiconvex so that
$$\mathcal{G}(u, \Omega) := \int_\Omega Qf(\nabla u)\, dx \to \operatorname{Min} \text{ in}$$
$$\mathcal{C} := \{u \in H^{1,p}(\Omega, \mathbb{R}^N) : u - u_0 \in \overset{\circ}{H}{}^{1,p}(\Omega, \mathbb{R}^N)\}$$
admits a solution for any given $u_0 \in H^{1,p}(\Omega, \mathbb{R}^N)$. Moreover,
$$\inf_{\mathcal{C}} \mathcal{F} = \inf_{\mathcal{C}} \mathcal{G}\,. \tag{1.5.15}$$

Theorem 1.5.7 *Let u denote the weak limit of any $\mathcal{F}$–minimizing sequence. Then $u \in C^{0,\mu}(\Omega, \mathbb{R}^N)$ and $\mathcal{L}^n(\{x \in \Omega : \nabla u(x) \text{ does not exist}\}) = 0$.*

Proof of Theorem 1.5.7: If $\{u_k\} \subset \mathcal{C}$ is an $\mathcal{F}$–minimizing sequence we have
$$\inf_{\mathcal{C}} \mathcal{F} = \inf_{\mathcal{C}} \mathcal{G} \leq \mathcal{G}(u_k, \Omega) \leq \mathcal{F}(u_k, \Omega) \xrightarrow[k\to\infty]{} \inf_{\mathcal{C}} \mathcal{F}$$
so that $\mathcal{G}(u_k, \Omega) \to \inf_{\mathcal{C}} \mathcal{G}$ and in conclusion $\mathcal{G}(u, \Omega) = \inf_{\mathcal{C}} \mathcal{G}$ by the lower semicontinuity of $\mathcal{G}$. Hence Theorem 1.5.6 applies to Qf, $\mathcal{G}$ in place of $f, \mathcal{F}$ and gives the result. $\square$

1.5 Variational Inequalities and Asymptotically Regular Integrands

Some ideas for the proof of Theorem 1.5.6:
(1) We assume $1 = \lim_{|A|\to\infty} |A|^{-p} f(A)$ which gives

$$\frac{1}{2}|A|^p - K \leq f(A) \leq \frac{3}{2}|A|^p + K$$

for some $K > 0$ depending on f. From this it easily follows that a Caccioppoli type inequality holds and that

$$\begin{cases} \nabla u \in L^s_{\text{loc}}(\Omega, \mathbb{R}^{nN}) \text{ for some } s > p \text{ and} \\ \left(\fint_{B_{R/2}} |\nabla u|^s \, dx\right)^{1/s} \leq c \cdot \left(\fint_{B_R} (1 + |\nabla u|)^p \, dx\right)^{1/p} \\ \text{for all balls } B_R \subset \Omega. \end{cases}$$

(2) For $\varepsilon > 0$ fixed consider

$$A_\varepsilon := \{(x, R) : B_R(x) \subset \Omega, \Phi(u, B_R(x)) \geq R^{p-\varepsilon}\},$$

$$\Phi(u, B_R(x)) := \fint_{B_R(x)} |u - (u)_R|^p \, dz$$

Proposition: *For each $\tau > 0$ there exists $R(\tau) > 0$ such that*

$$\Phi(u, B_{\tau R}(x)) \leq 2 \cdot C_0 \tau^p \, \Phi(u, B_R(x))$$

holds for all $(x, R) \in A_\varepsilon$, $R \leq R(\tau)$.

Here C_0 denotes the absolute constant appearing in Uhlenbeck's estimate

$$\fint_{B_r(x)} |h - (h)_r|^p \, dx \leq C_0 \left(\frac{r}{R}\right)^p \fint_{B_R(x)} |h - (h)_R|^p \, dx$$

for solutions h of $\partial_\alpha(|\nabla h|^{p-2} \partial_\alpha h) = 0$.

If the statement of the Proposition were wrong then there are $\tau \in (0,1)$ and sequences $(x_\nu, R_\nu) \in A_\varepsilon$, $R_\nu \to 0$, such that

$$\Phi(u, B_{\tau R_\nu}(x_\nu)) > 2 C_0 \, \tau^p \, \Phi(u, B_{R_\nu}(x_\nu)).$$

Let

$$\delta_\nu := \Phi(u, B_{R_\nu}(x_\nu))^{1/p}$$

and

$$u_\nu(z) := \delta_\nu^{-1} [u(x_\nu + R_\nu z) - (u)_{R_\nu, x_\nu}], \quad z \in B_1.$$

These functions satisfy $\Phi(u_\nu, B_\tau) > 2 C_0 \cdot \tau^p$. Letting

$$f_\nu(A) := \left(\frac{R_\nu}{\delta_\nu}\right)^p \cdot f(\delta_\nu \cdot R_\nu^{-1} A)$$

and

$$\mathcal{F}_\nu(w, B_1) := \int_{B_1} f_\nu(\nabla w)\, dx$$

one easily checks

$$\mathcal{F}\big(u, B_{R_\nu}(x_\nu)\big) = \delta_\nu^p\, R_\nu^{-p+n}\, \mathcal{F}_\nu(u_\nu, B_1)$$

so that u_ν is locally $\mathcal{F}_\nu(\cdot, B_1)$-minimizing. After passing to subsequences we have $u_\nu \rightharpoonup u_0$ in $H^{1,p}_{\text{loc}}(B_1, \mathbb{R}^N)$ (observe $\fint_{B_1} |u_\nu|^p\, dx = 1$, Caccioppoli's inequality gives local gradient bounds) and (on account of $\fint_{B_1} |u_\nu|^p\, dx = 1$) $u_\nu \to u_0$ weakly in $L^p(B_1, \mathbb{R}^N)$ (and strongly in $L^p_{\text{loc}}(B_1)$). Using (1) as well as $|A|^{-p} f(A) \to 1$, $|A| \to \infty$, one now can prove that u_0 is a local minimizer of $\int_{B_1} |\nabla w|^p\, dx$, hence

$$\fint_{B_\tau} |u_0 - (u_0)_\tau|^p\, dx \le C_0 \left(\frac{\tau}{R}\right)^p \fint_{B_R} |u_0 - (u_0)_R|^p\, dx$$

for any $R \in (\tau, 1)$. Since u_0 is in $L^p(B_1, \mathbb{R}^N)$ we may pass to the limit $R \nearrow 1$ so that

$$\fint_{B_\tau} |u_0 - (u_0)_\tau|^p\, dx \le C_0 \tau^p \fint_{B_1} |u_0|^p\, dx$$

$$\le C_0 \tau^p \liminf_{\nu \to \infty} \fint_{B_1} |u_\nu|^p\, dx \le C_0 \tau^p.$$

For $\nu \gg 1$ we then have $\Phi(u_\nu, B_\tau) \le \frac{3}{2} C_0 \tau^p$ which is a contradiction.

(3) Suppose $\mu \in (0, 1)$ is given and define $\varepsilon = p \cdot (1 - \mu)$. For τ being determined later choose $R(\tau)$ as in (2) and let $r := \min\big(R(\tau), \frac{1}{2} \text{dist}(x, \partial\Omega)\big)$ where $x \in \Omega$ is arbitrary. Finally let $r_k = \tau^k \cdot r$.

Case 1: $\Phi\big(u, B_{r_{k-1}}(x)\big) \le r_{k-1}^\alpha$, $\alpha := p - \varepsilon$. Then $\Phi\big(u, B_{r_k}(x)\big) \le c_1 \cdot \tau^{-n-\alpha}\, r_k^\alpha$.

Case 2: $\Phi\big(u, B_{r_{k-1}}(x)\big) > r_{k-1}^\alpha$. Then we deduce from the Proposition ($\Theta := 2 C_0\, \tau^p$)

$$\Phi\big(u, B_{r_k}(x)\big) \le \Theta \Phi\big(u, B_{r_{k-1}}(x)\big)$$

so that in both cases

$$\Phi\big(u, B_{r_k}(x)\big) \le \max\big(c_1\, \tau^{-n-\alpha}\, r_k^\alpha,\ \Theta \cdot \Phi\big(u, B_{r_{k-1}}(x)\big)\big)$$

1.5 Variational Inequalities and Asymptotically Regular Integrands

which gives by iteration and after choosing

$$\tau := \left(\frac{1}{4C_0}\right)^{1/\varepsilon}:$$

$$\Phi(u, B_{r_k}(x)) \leq c_1 \tau^{\alpha \cdot k} \max\left(r^\alpha \cdot \tau^{-n-\alpha}, c_1^{-1} 2^{-k} \cdot \Phi(u, B_r(x))\right)$$

$$\leq c_2 \tau^{\alpha \cdot k} \max\left(r^\alpha, \Phi(u, B_r(x))\right).$$

For $\rho < r$ we infer (choose k such that $r_k \leq \rho < r_{k-1}$)

$$\Phi(u, B_\rho(x)) \leq c_3 \left(\frac{\rho}{r}\right)^\alpha \cdot \max(r, \Phi(r))$$

$$= c_3 \left(\frac{\rho}{r}\right)^{p \cdot \mu} \max(r, \Phi(r)),$$

hence $u \in C^{0,\mu}(\Omega, \mathbb{R}^N)$.

(4) Let $\Omega^* := \{y \in \Omega : \sup_{r>0} \fint_{B_r(y)} |\nabla u|^p \, dx < \infty\}$.

Proposition: *For any $x \in \Omega^*$ there are constants $K_x > 0$ and $R_x < \frac{1}{2} \operatorname{dist}(x, \partial\Omega)$ as follows: if for some $R \leq R_x$, $\operatorname*{osc}_{B_R(x)} u \geq K_x \cdot R$ then $\operatorname*{osc}_{B_{R/2}(x)} u \leq \frac{1}{2} \operatorname*{osc}_{B_R(x)} u$.*

Again we argue by contradiction: Suppose that $0 \in \Omega^*$ and that there are sequences $R_\nu \to 0$, $K_\nu \to \infty$ such that

$$\omega_\nu := \operatorname*{osc}_{B_{R_\nu}} u \geq K_\nu R_\nu$$

and

$$\operatorname*{osc}_{B_{R_\nu/2}} u > \frac{1}{2} \omega_\nu.$$

We then scale

$$u_\nu(z) := \frac{1}{\omega_\nu} [u(R_\nu z) - (u)_{R_\nu}]$$

and obtain a weak limit v which on account of $0 \in \Omega^*$ is seen to be constant, hence $\operatorname*{osc}_{B_{1/2}} v = 0$. On the other hand since u_ν are local minimizers of suitable functionals we have uniform bounds on $[u_\nu]_{C^{0,\mu}}$ so that $\operatorname*{osc}_{B_{1/2}} u_\nu \to \operatorname*{osc}_{B_{1/2}} v = 0$. But by assumption we have $\operatorname*{osc}_{B_{1/2}} u_\nu = \frac{1}{\omega_\nu} \cdot \operatorname*{osc}_{B_{R_\nu/2}} u > \frac{1}{2}$. □

Now let $x \in \Omega^*$ denote an arbitrary point and calculate $K = K_x$, $R = R_x$ according to the proposition.

Case 1: $\underset{B_R(x)}{\operatorname{osc}} u \leq K \cdot R$. Then also $\underset{B_{R/2}(x)}{\operatorname{osc}} u \leq K \cdot R$.

Case 2: $\underset{B_R(x)}{\operatorname{osc}} u \geq K \cdot R \implies \underset{B_{R/2}(x)}{\operatorname{osc}} u \leq \frac{1}{2} \underset{B_R(x)}{\operatorname{osc}} u$.

In both cases $\underset{B_{R/2}(x)}{\operatorname{osc}} u \leq \max(K \cdot R, \frac{1}{2} \underset{B_R(x)}{\operatorname{osc}} u)$, hence

$$\underset{B_{2^{-k} \cdot R}(x)}{\operatorname{osc}} u \leq \max(2^{-k+1} \cdot K \cdot R, 2^{-k} \underset{B_R(x)}{\operatorname{osc}} u)$$

so that $\underset{B_\rho(x)}{\operatorname{osc}} u \leq \max(4 \cdot K, 2 \cdot R^{-1} \underset{B_R(x)}{\operatorname{osc}} u) \cdot \rho$, $\rho \leq R$.

We deduce $\underset{\Omega \ni y \to x}{\lim\sup} |u(y) - u(x)|/|x-y| < \infty$ for all $x \in \Omega^*$ and by Stepanoff's theorem $\nabla u(x)$ exists for $\mathcal{L}^n$ – almost all $x \in \Omega^*$. Since clearly $\mathcal{L}^n(\Omega - \Omega^*) = 0$ the proof is complete. □

1.6 Approximations for some Model Problems in Nonlinear Elasticity

In nonlinear elasticity the physical behaviour of so-called hyperelastic materials can be characterized with the help of variational methods (see [4], [5] for details): suppose that in the undeformed state the body occupies a region $\Omega \subset \mathbb{R}^n$ (usually $n = 2$ or $n = 3$). Then we look for minimizers $u : \overline{\Omega} \to \mathbb{R}^n$ of the stored energy $I(u) = \int_\Omega W(\nabla u) dx$ in admissible classes $\mathcal{C}$ of deformations which are required to be locally orientation preserving, i.e. $\det \nabla u(x) > 0$ for $x \in \Omega$. The stored energy density W (characterizing the mechanical properties of the material) is a function of all real $(n \times n)$-matrices F with nonnegative determinants (for simplicity we consider homogeneous materials) which has to satisfy certain growth conditions and also a nondegeneracy condition of the form

$$\lim_{k \to \infty} W(F_k) = \infty \quad \text{if} \quad \lim_{k \to \infty} \det F_k = 0 \tag{1.6.1}$$

saying that an infinite amount of energy is needed to shrink a finite volume to zero. Under the assumption that W is in addition a polyconvex function Ball (compare [4]) proved the existence of solutions u to the minimum problem

$$I \to \min \quad \text{in } \mathcal{C} \tag{1.6.2}$$

1.6 Approximations for some Model Problems in Nonlinear Elasticity

provided $\mathcal{C}$ is embedded in some suitable Sobolev space. Due to the sign condition imposed on the Jacobian of all admissible comparison functions it is by no means obvious that a minimizer u is also a solution (at least in the sense of distributions) of the corresponding equilibrium equation

$$\operatorname{div} \frac{\partial W}{\partial F}(\nabla u) = 0 \quad \text{on } \Omega. \tag{1.6.3}$$

In fact, the standard variation $u_t(x) = u(x) + t\rho(x)$, $\rho \in C_0^\infty(\Omega, \mathbb{R}^n)$, may violate the side condition $\det \nabla u_t > 0$ on a subset of Ω with positive measure. At the same time the regularity theory concerning problem (1.6.2) under natural hypotheses imposed on the integrand W is very poor for the reason that all techniques which we previously applied to vectorvalued problems do not take care of $\det \nabla u > 0$. To my knowledge the only results concern the case $\mathcal{C} \subset H^{1,p}(\Omega, \mathbb{R}^n)$ for some $p \geq n$. If $p > n$ then continuity follows from Sobolev's embedding, the limit case $p = n$ is treated in [79] and the arguments used there do not rely on the minimality of the functions under consideration. So in general it is unknown under what conditions a minimizer is regular (of class C^1) and in addition provides us with a solution of (1.6.3).

In this section we want to describe some ideas which might lead to (partial) regularity results for problem (1.6.2) and at the same time serve as a basis for numerical computations. The results are taken from the papers [40, 41, 42].

First we limit ourselves to the two-dimensional case $n = 2$ and consider stored energy densities W of the form

$$W(F) = \frac{1}{2}|F|^2 + h(\det F) \tag{1.6.4}$$

for a nonnegative function h defined on the positive real numbers. In order to have the limit behaviour (1.6.1) it would be necessary to require $h(0) = \infty$. Unfortunately our discussion covers only the case that $h(0)$ is an arbitrary but finite value and we then will prove that any minimizer of $\int_\Omega W(\nabla u)dx$ under Dirichlet boundary conditions and the constraint $\det \nabla u(x) \geq 0$ is Hölder continuous on the domain Ω. But even in this case it is an open problem if (1.6.3) holds in the weak sense.

Theorem 1.6.1 *Suppose that $\Omega \subset \mathbb{R}^2$ is a bounded domain and let $h : [0, \infty) \to [0, \infty)$ denote a function of class C^1 such that $|h(t) - th'(t)|$ is bounded independent of $t \geq 0$. Consider a function $u \in H^{1,1}(\Omega, \mathbb{R}^2)$ such that*

$$I(u) = \int_\Omega W(\nabla u)dx < \infty,$$

$$\det \nabla u(x) \geq 0 \quad \text{for a.a. } x \in \Omega,$$

$$\operatorname{div}\left((\nabla u)^T \frac{\partial W}{\partial F}(\nabla u) - W(\nabla u)\mathbf{1}\right) = 0 \qquad (1.6.5)$$

in the sense of distributions

where W is defined in (1.6.4) and $\mathbf{1}$ denotes the identity. Then we have $u \in C^{0,\gamma}(\Omega, \mathbb{R}^2)$ for any $\gamma \in (0,1)$.

Remarks:

1. Equation (1.6.5) states
$$0 = \frac{d}{dt/0} \int_\Omega W(\nabla u_t) dx$$
for variations u_t of the form $u_t(x) = u(x + t\rho(x))$, $\rho \in C_0^\infty(\Omega, \mathbb{R}^2)$.

2. We want to describe a natural setting for which all assumptions of Theorem 1.6.1 are satisfied: Suppose that $\partial\Omega$ is Lipschitz and fix an open part Γ of $\partial\Omega$ with positive measure on which a function u_0 is given. We then consider the class $\mathcal{C} = \{u \in H^{1,1}(\Omega, \mathbb{R}^2) : u|_\Gamma = u_0, \det \nabla u \geq 0 \text{ a.e.}\}$ which is required to contain a function v with the property $\int_\Omega W(\nabla v) dx < \infty$. Suppose further that $h: [0, \infty) \to [0, \infty)$ is a convex C^1-function such that $\sup\{|h(t) - th'(t)| : t \geq 0\} < \infty$, for example we may take $h(t) = t + (t+\varepsilon)^{-s}$, $\varepsilon, s > 0$. Then there exists $u \in \mathcal{C}$ with the property $I(u) \leq I(w)$ for all $w \in \mathcal{C}$, especially $\frac{d}{dt/0} I(u_t) = 0$ since $u_t(x) = u(x + t\rho(x))$, $\rho \in C_0^\infty(\Omega, \mathbb{R}^2)$, belongs the class $\mathcal{C}$ and thereby is an admissible variation.

3. It would be interesting to know if our result can be improved to Lipschitz regularity.

4. Slightly more general energy densities are discussed in [40] where also an additional side condition of the form $u(\Omega) \subset \overline{M}$ for a bounded region $M \subset \mathbb{R}^2$ is included.

Proof of Theorem 1.6.1: Let
$$G(F) = F^T F - \frac{1}{2}|F|^2 \mathbf{1}, \quad F \in \mathbb{R}^{2\times 2},$$
$$p(t) = h(t) - th'(t), \quad t \geq 0,$$

and observe that $G(F)$ is symmetric with vanishing trace so that we may write

$$G(F) = \begin{pmatrix} G_{11}(F) & G_{12}(F) \\ G_{12}(F) & -G_{11}(F) \end{pmatrix}$$

1.6 Approximations for some Model Problems in Nonlinear Elasticity

and equation (1.6.5) reads

$$\operatorname{div} G(\nabla u) = \nabla(p(\det \nabla u)) \tag{1.6.6}$$

where the divergence operator has to be applied linewise. We first show:

Lemma 1.6.1 *Under the assumptions of Theorem 1.6.1 we have* $G(\nabla u) \in L^s_{\text{loc}}(\Omega, \mathbb{R}^{2\times 2})$ *for any* $s < \infty$.

For small $\varepsilon > 0$ let $\tilde{G}_\varepsilon$ and $\tilde{p}_\varepsilon$ denote regularizations of the functions $G(\nabla u)$ and $p(\det \nabla u)$. Then (1.6.6) implies

$$\operatorname{div} \tilde{G}_\varepsilon(x) = \nabla \tilde{p}_\varepsilon(x), \quad x \in \Omega, \ \operatorname{dist}(x, \partial\Omega) > \varepsilon. \tag{1.6.7}$$

Consider a disc $B_R(x_0) = \{z \in \mathbb{R}^2 : |z - x_0| < R\}$ contained in Ω such that $\operatorname{dist}(B_R(x_0), \partial\Omega) > 0$. We multiply both sides of (1.6.7) with the vector $Q(x - x_0)$, $Q \in \mathbb{R}^{2\times 2}$, and obtain after integration over $B_R(x_0)$ the relation

$$\int_{B_R(x_0)} Q(x - x_0) \cdot \operatorname{div} \tilde{G}_\varepsilon(x) dx \tag{1.6.8}$$

$$= \int_{\partial B_R(x_0)} (\tilde{G}_\varepsilon(x)\nu(x)) \cdot Q(x-x_0) d\mathcal{H}^1(x) - \int_{B_R(x_0)} \tilde{G}_\varepsilon(x) Q dx$$

$$= \int_{\partial B_R(x_0)} \tilde{p}_\varepsilon(x)\nu(x) \cdot Q(x-x_0) d\mathcal{H}^1(x) - \int_{B_R(x_0)} \tilde{p}_\varepsilon(x) \operatorname{trace} Q \, dx.$$

Here $\nu(x) = \frac{1}{R}(x - x_0)$. Let us introduce the kernels

$$K_1(z) = |z|^{-4}(z_1^2 - z_2^2), \quad K_2(z) = |z|^{-4} 2z_1 z_2$$

for $z = (z_1, z_2) \in \mathbb{R}^2 - \{0\}$. If we choose $Q = \begin{pmatrix} 1 & 0 \\ 0 & -1 \end{pmatrix}$ then (1.6.8) implies for a.a. R as above

$$\int_{\partial B_R(x_0)} |x-x_0|^{-1}\{(x_1 - x_{01})^2 - (x_2 - x_{02})^2\} \tilde{p}_\varepsilon(x) d\mathcal{H}^1(x)$$

$$= R\int_{\partial B_R(x_0)} (\tilde{G}_\varepsilon(x))_{11} d\mathcal{H}^1(x) - 2\int_{B_R(x_0)} (\tilde{G}_\varepsilon(x))_{11} dx \tag{1.6.9}$$

$$= \pi R^3 \frac{d}{dR}\int_{B_R(x_0)} (\tilde{G}_\varepsilon(x))_{11} dx.$$

Next we integrate (1.6.9) from r to R and get

$$\fint_{B_R(x_0)} (\tilde{G}_\varepsilon(x))_{11} dx - \fint_{B_r(x_0)} (\tilde{G}_\varepsilon(x))_{11} dx$$

$$= \frac{1}{\pi} \fint_{B_R(x_0) - B_r(x_0)} K_1(x-x_0)\tilde{p}_\varepsilon(x)dx.$$

After passing to the limit $\varepsilon \downarrow 0$ we obtain

$$\fint_{B_R(x_0)} G_{11}(\nabla u)(x)dx - \fint_{B_r(x_0)} G_{11}(\nabla u)(x)\, dx = T_r(\tilde{p})(x_0) - T_R(\tilde{p})(x_0) \tag{1.6.10}$$

for all $B_r(x_0) \subset B_R(x_0) \subset \Omega$. In (1.6.10) we use the notation

$$\tilde{p}(x) = \begin{cases} p(\det \nabla u(x)), & x \in \Omega \\ 0, & x \in \mathbb{R}^2 - \Omega \end{cases},$$

$$T_R(\tilde{p})(x_0) = \frac{1}{\pi} \int_{\mathbb{R}^2 - B_R(x_0)} \tilde{p}(x)\, K_1(x-x_0)\, dx.$$

Recalling $\tilde{p} \in L^\infty(\mathbb{R}^2)$ standard arguments from singular integral theory (see [77]) imply

$$T_r(\tilde{p}) \to T(\tilde{p}) \quad \text{in } L^t(\mathbb{R}^2)$$

for any finite t where $T(\tilde{p})(x_0) = \frac{1}{\pi} \int_{\mathbb{R}^2} K_1(x-x_0)\tilde{p}(x)dx$. Passing to the limit $r \downarrow 0$ in (1.6.10) we thus deduce

$$\fint_{B_R(x_0)} G_{11}(\nabla u)(x)dx - G_{11}(\nabla u)(x_0) = T(\tilde{p})(x_0) - T_R(\tilde{p})(x_0)$$

being valid for a.a. $x_0 \in \Omega_R = \{y \in \Omega : \operatorname{dist}(y, \partial\Omega) > R\}$. This clearly gives $G_{11}(\nabla u) \in L^t_{\mathrm{loc}}(\Omega)$. Using $Q = \begin{pmatrix} 0 & 1 \\ 1 & 0 \end{pmatrix}$ in (1.6.8) $G_{12}(\nabla u) \in L^t_{\mathrm{loc}}(\Omega)$ is proved in a similar way. This finishes the proof of Lemma 1.6.1.

We now continue with the *Proof of the Theorem:* From $\det \nabla u \geq 0$ a.e. we infer

$$\frac{1}{\sqrt{2}} |\nabla u|^2 \leq |G(\nabla u)| + \sqrt{2} \det \nabla u,$$

hence

$$\int_{B_r(x_0)} |\nabla u|^2\, dx \leq \sqrt{2} \int_{B_r(x_0)} |G(\nabla u)|dx + 2 \int_{B_r(x_0)} \det \nabla u\, dx.$$

The last integral involving $\det \nabla u$ is handled with the help of the following isoperimetric inequality (compare [70])

1.6 Approximations for some Model Problems in Nonlinear Elasticity

$$\int_{B_r(x_0)} \det \nabla u \, dx \leq \frac{r}{2} \int_{\partial B_r(x_0)} [\nabla u]^2(x) d\mathcal{H}^1(x)$$

being valid for almost all $r > 0$. Here $[F] := \sup_{|a|=1} |Fa|$ denotes the operator norm of $F \in \mathbb{R}^{2 \times 2}$. Clearly

$$[F]^2 = \frac{1}{2}|F|^2 + \sqrt{2}|G(F)|$$

so that we arrive at

$$\int_{B_r(x_0)} |\nabla u|^2 dx \leq \frac{r}{2} \int_{\partial B_r(x_0)} |\nabla u|^2 d\mathcal{H}^1(x) + \sqrt{2} \int_{B_r(x_0)} |G(\nabla u)| \, dx$$

$$+ \frac{r}{\sqrt{2}} \int_{\partial B_r(x_0)} |G(\nabla u)| d\mathcal{H}^1(x).$$

In a final step we rewrite this result as differential inequality for the quantity $r^{-2} \int_{B_r(x_0)} |\nabla u|^2 dx$ and use Lemma 1.6.1 to estimate the terms involving $G(\nabla u)$. The claim then follows from Morrey's lemma. □

We next discuss a different way of approximating problems in nonlinear elasticity. To be precise consider a bounded open set $\Omega \subset \mathbb{R}^n$, $n \geq 2$, representing the undeformed state of the body and fix some real number $p \geq n$. Suppose further that we are given a function $u_0 \in H^{1,p}(\Omega, \mathbb{R}^n)$ such that

$$\tau \leq \det \nabla u_0(x) \leq \frac{1}{\tau} \quad \text{a.e. on } \Omega$$

for some $\tau \in (0, 1)$. Then we look at the variational problem

$$J(u) = \int_\Omega |\nabla u|^p + H(\det \nabla u) \, dx \to \min \quad (1.6.11)$$
$$\text{in } \mathcal{C} := \{w \in H^{1,p}(\Omega, \mathbb{R}^n) : w = u_0 \text{ on } \partial\Omega\}$$

with $H : (0, \infty) \to [0, \infty)$ of class C^2, strictly convex and with the property

$$\lim_{t \downarrow 0} H(t) = \infty. \quad (1.6.12)$$

We define $H(t) = \infty$ for $t \leq 0$. From the work of Ball [4] we deduce the existence of solutions to (1.6.11) (compare also Theorem 1.6.2) which belong to the class $C^0(\Omega, \mathbb{R}^n)$. As mentioned before nothing is known about C^1-regularity of minimizers but the results described below give rise to the following

Conjecture: *Suppose that $u \in \mathcal{C}$ is a solution of (1.6.11) under the natural growth hypothesis (1.6.12). Then there is an open subset Ω_0 of Ω whose complement has vanishing Lebesgue measure such that $u \in C^{1,\alpha}(\Omega_0)$ for any $0 < \alpha < 1$. Moreover, $x_0 \in \Omega_0$ if and only if the following conditions hold:*

a) *x_0 is a Lebesgue point for ∇u*

b) *$\det \nabla u(x_0) > 0$*

c) *$\lim\limits_{r \downarrow 0} \fint_{B_r(x_0)} |\nabla u - (\nabla u)_{x_0,r}|^p dx = 0$.*

As a first approach towards the conjecture we consider the case

$$\lim_{t \to \infty} H'(t) = \infty$$

and replace (1.6.11) by a sequence of more regular problems

$$J_\delta(v) = \int_\Omega |\nabla v|^p + H_\delta(\det \nabla v) dx \to \min \text{ in } \mathcal{C} \qquad (1.6.13)$$

where for $0 < \delta < \tau$

$$H_\delta(t) = \begin{cases} H'(\delta)(t - \delta) + H(\delta), & t \leq \delta \\ H(t), & \delta \leq t \leq \delta^{-1} \\ H'(\frac{1}{\delta})(t - \frac{1}{\delta}) + H(\frac{1}{\delta}), & t \geq \delta^{-1} \end{cases}$$

is defined for all $t \in \mathbb{R}$ with linear growth at $\pm\infty$. It is easy to show that

$$H_\varepsilon(t) \leq H_\delta(t) \quad \text{if } \delta \leq \varepsilon, \quad H_\delta(t) \to H(t) \quad \text{as } \delta \downarrow 0,$$

and in [41] we used these properties to obtain

Theorem 1.6.2 *Problem (1.6.13) admits a solution u_δ which is of class $C^{0,\gamma}(\Omega)$ for all $\gamma < 1$. After passing to a subsequence we also have $u_\delta \to u$ ($\delta \downarrow 0$) strongly in $H^{1,p}(\Omega, \mathbb{R}^n)$ for some function $u \in \mathcal{C}$ and u is a solution of (1.6.11).*

It is worth remarking that Theorem 1.6.2 gives existence of solutions to (1.6.11) without using the elaborate arguments of Ball [4].

As a first step towards our conjecture we now discuss the regularity properties of the minimizers u_δ.

Theorem 1.6.3 *There exists an open subset Ω_δ of Ω such that $u_\delta \in C^{1,\alpha}(\Omega_\delta)$ for any $0 < \alpha < 1$. We have the estimates*

$$\mathcal{L}^n(\Omega - \Omega_\delta) \leq \min\left\{H(\delta), H(\frac{1}{\delta})\right\}^{-1} J(u_0) \xrightarrow[\delta \downarrow 0]{} 0$$

and $\delta < \det \nabla u_\delta(x) < \delta^{-1}$ on Ω_δ.

1.6 Approximations for some Model Problems in Nonlinear Elasticity

Proof of Theorem 1.6.3: We fix $\delta \in (0, \tau)$ and write u and H in place of u_δ and H_δ. Suppose that we are given $A_0 \in \mathbb{R}^{n \times n}$ such that

$$a_0 := \det A_0 \in \left(\delta, \frac{1}{\delta}\right).$$

Then we can calculate $\sigma = \sigma(A_0, \delta)$ such that

$$\det A \in \left[\frac{1}{2}(a_0 + \delta), \frac{1}{2}\left(a_0 + \frac{1}{\delta}\right)\right] \subset \left(\delta, \frac{1}{\delta}\right)$$

holds for all $A \in \mathbb{R}^{n \times n}$, $|A - A_0| \leq \sigma$.

Lemma 1.6.2 *There is a constant $c_* = c_*(A_0, p, H''(a_0))$ with the following property: For each $t \in (0, 1)$ there exists $\varepsilon = \varepsilon(A_0, t, \delta)$ such that, for every ball $B_R(x_0) \subset \Omega$, the conditions*

$$|(\nabla u)_{x_0, R} - A_0| \leq \sigma,$$

$$E(u, B_R(x_0)) := \fint_{B_R(x_0)} |\nabla u - (\nabla u)_{x_0, R}|^2 + |\nabla u - (\nabla u)_{x_0, R}|^p dx < \varepsilon^2$$

imply

$$E(u, B_{tR}(x_0)) \leq c_* t^2 E(u, B_R(x_0)).$$

From this result the statement of Theorem 1.6.3 follows in a routine manner: Let $\Omega'_\delta = \{x \in \Omega : \delta < \det \nabla u(x) < \delta^{-1}\}$. Minimality of u implies

$$\int_{\Omega - \Omega'_\delta} H(\det \nabla u) dx \leq J(u_0)$$

and the term on the left is bounded below by the quantity

$$\mathcal{L}^n(\Omega - \Omega'_\delta) \min\left\{H(\delta), H\left(\frac{1}{\delta}\right)\right\}.$$

Next consider $x_0 \in \Omega'_\delta$ such that

x_0 is a Lebesgue point for ∇u,

$$\delta < \det \nabla u(x_0) < \delta^{-1}, \tag{1.6.14}$$

$$E(u, B_r(x_0)) \to 0 \text{ as } r \downarrow 0.$$

Clearly these conditions are true for almost all $x_0 \in \Omega'_\delta$. If we take

$$A_0 = \lim_{r \downarrow 0}(\nabla u)_{x_0, r} \quad (= \nabla u(x_0))$$

then iteration of Lemma 1.6.2 gives $u \in C^1$ in a neighborhood of x_0. This shows $u \in C^1(\Omega_\delta)$ with

$$\Omega_\delta = \{x_0 \in \Omega'_\delta : x_0 \text{ satisfies } (1.6.14)\}$$

and the proof of Theorem 1.6.3 is complete.

The *Proof of Lemma 1.6.2* proceeds in several steps: We fix $t \in (0,1)$ and define c_* later on. If the lemma were false then we find a sequence of balls $B_{R_k}(x_k) \subset \Omega$ such that

$$|A_k - A_0| \leq \sigma, \quad A_k := (\nabla u)_{x_k, R_k},$$

$$E(u, B_{R_k}(x_k)) = \varepsilon_k^2 \to 0$$

but

$$E(u, B_{tR_k}(x_k)) > c_* t^2 E(u, B_{R_k}(x_k)).$$

We let

$$v_k(z) = \varepsilon_k^{-1} R_k^{-1}[u(x_k + R_k z) - (u)_{x_k, R_k} - R_k A_k(z)], \quad z \in B_1;$$

at least for a subsequence we can arrange

$$A_k \to: A, \quad \det A \in \left[\frac{1}{2}(a_0 + \delta), \frac{1}{2}(a_0 + \delta^{-1})\right],$$

$$v_k \to: v \quad \text{weakly in } H^{1,2}(B_1, \mathbb{R}^n),$$

$$\varepsilon_k^{1-2/p} \nabla v_k \to 0 \quad \text{weakly in } L^p(B_1, \mathbb{R}^{n \times n}).$$

After some calculations we deduce the following limit equation satisfied by v:

$$\int_{B_1} p|A|^{p-2}[\nabla v + (p-2)|A|^{-2}(A \cdot \nabla v)A] \cdot \nabla \varphi$$

$$+ H''(\det A)(\operatorname{Cof} A \cdot \nabla v)(\operatorname{Cof} A \cdot \nabla \varphi) \, dx = 0 \quad (1.6.15)$$

being valid for all $\varphi \in C_0^1(B_1, \mathbb{R}^n)$. Here Cof A denotes the cofactor matrix of A which by definition satisfies

$$A \operatorname{Cof} A = \det A \, \mathbf{1}.$$

Since $A \neq 0$ and $H''(\det A) \geq 0$ (1.6.15) is a linear elliptic system with constant coefficients. This gives $v \in C^\infty(B_1)$ and the estimate

$$\fint_{B_t} |\nabla v - (\nabla v)_t|^2 dx \leq c_0 t^2 \fint_{B_1} |\nabla v - (\nabla v)_1|^2 \, dx,$$

1.6 Approximations for some Model Problems in Nonlinear Elasticity

$c_0 = c_0\left(A_0, n, p, C^2\text{-norms of } H \text{ near det } A_0\right)$. We define $c_* = 2c_0$. From our assumption we deduce

$$\fint_{B_t} |\nabla v_k - (\nabla v_k)_t|^2 + \varepsilon_k^{p-2}|\nabla v_k - (\nabla v_k)_t|^p \, dx > c_* t^2.$$

On the other hand – using convexity of H as well as smoothness of the limit function v – one can in fact show that

$$\nabla v_k \to \nabla v \quad \text{in } L^2_{\text{loc}}(B_1, \mathbb{R}^{n \times n}),$$

$$\varepsilon_k^{1-2/p} \nabla v_k \to 0 \quad \text{in } L^p_{\text{loc}}(B_1, \mathbb{R}^{n \times n}),$$

hence

$$\fint_{B_t} |\nabla v - (\nabla v)_t|^2 dx \geq c_* t^2.$$

From weak convergence in $H^{1,2}(B_1, \mathbb{R}^n)$ we infer

$$\fint_{B_1} |\nabla v|^2 dx \leq \lim_{k \to \infty} \fint_{B_1} |\nabla v_k|^2 dx \leq 1,$$

$$\fint_{B_1} \nabla v \, dx = 0$$

so that

$$\fint_{B_t} |\nabla v - (\nabla v)_t|^2 dx \geq c_* t^2 \fint_{B_1} |\nabla v - (\nabla v)_1|^2 dx$$

which is the desired contradiction. □

2 Manifolds of Prescribed Mean Curvature in the Setting of Geometric Measure Theory

2.1 The Mean Curvature Problem

In these lectures we want to show how the powerful techniques of Geometric Measure Theory can be applied to solve the following

Mean Curvature Problem: (MCP)

> Given a $(m-1)$-dimensional submanifold Γ of Euclidean space $\mathbb{R}^{m+k}$, find some m-dimensional manifold Σ such that Σ has boundary Γ and that the mean curvature of Σ is prescribed.

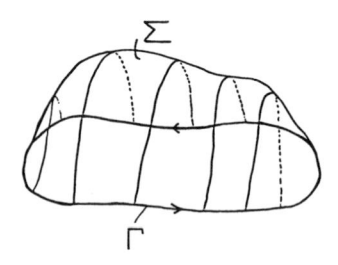

Γ plays the role of a prescribed boundary so that MCP in the form stated above is the generalization of classical Plateau's problem for surfaces of prescribed mean curvature in Euclidean space $\mathbb{R}^3$ to the case of arbitrary dimension m (≥ 2) and arbitrary codimension k (≥ 1). For zero curvature we also include the minimal surface case.

Before going into the history of the MCP and related geometric questions let us give a precise meaning to the notion of mean curvature. The basic reference is [74].

Let Σ denote a smoth m-dimensional submanifold of $\mathbb{R}^{m+k}$. Fix some point $y \in \Sigma$ and define for $\tau \in T_y\Sigma$

$$D_\tau f := \frac{d}{dt/0} f(\gamma(t)),$$

γ a C^1-curve in Σ s.t. $\gamma(0) = y$, $\dot\gamma(0) = \tau$

2.1 The Mean Curvature Problem

as the *directional derivative* of a function $f : \Sigma \to \mathbb{R}^\ell$ (at y in the direction τ).

For real valued f let

$$\nabla^\Sigma f(y) := \sum_{j=1}^{m} (D_{\tau_j} f) \tau_j$$

for any ONB $\{\tau_j\}$ of $T_j\Sigma$. $\nabla^\Sigma f$ is the *relative gradient* of f at y, clearly

$$\nabla^\Sigma f(x) = \text{Projection of } \nabla f(y) \text{ onto } T_y\Sigma$$

if f is smooth on a full neighborhood of y in $\mathbb{R}^{m+k}$.

Consider a vectorfield $X : \Sigma \to \mathbb{R}^{m+k}$, $X = (X^1, \ldots, X^{m+k})$. We define the *relative divergence of X* as

$$\begin{aligned}\operatorname{div}_\Sigma X &= \sum_{i=1}^{m+k} e_i \cdot (\nabla^M X^i) \\ &= \sum_{j=1}^{m} (D_{\tau_j} X) \cdot \tau_j\end{aligned}$$

where $e_1, \ldots, e_{m+k}$ is the standard basis in $\mathbb{R}^{m+k}$ and $\tau_1, \ldots, \tau_m$ denotes an orthonormal basis in $T_y\Sigma$.

Finally we introduce the *2nd fundamental form of Σ at y* as

$$B_y : T_y\Sigma \times T_y\Sigma \longrightarrow (T_y\Sigma)^\perp$$

$$B_y(\tau, \eta) := -\sum_{\alpha=1}^{k} (\eta \cdot D_\tau \nu^\alpha) \nu^\alpha(y)$$

where $\nu^1, \ldots, \nu^k$ are locally defined (near y) vectorfields with the property

$$\begin{cases} \nu^\alpha \cdot \nu^\beta \equiv \delta_{\alpha\beta}, \\ \nu^\alpha(z) \in T_\Sigma(z)^\perp \end{cases}$$

for points z in a neighborhood of y in Σ.

It is well known that B_y is a symmetric bilinear form with values in the normal space $T_y\Sigma^\perp$.

Definition (Mean curvature vector): The geometric mean curvature vector $\underline{H}(y)$ of Σ at y is the quantity

$$\underline{H}(y) = \sum_{i=1}^{m} B_y(\tau_i, \tau_i)$$

for any ONB $\{\tau_i\}$ of $T_y\Sigma$.

Remarks:

1. Note that we did not impose orientability on Σ.

2. The geometric mean curvature vector is just the trace of the second fundamental form. Many authors use $\frac{1}{m}\sum_{i=1}^{m} B_y(\tau_i, \tau_i)$ which sometimes leads to some confusion.

3. We have the formula

$$\boxed{\underline{H} = -\sum_{i=1}^{k} \left(\operatorname{div}_\Sigma \nu^\alpha\right)\nu^\alpha}$$

for the mean curvature vector. Here $\{\nu^\alpha\}_{\alpha=1,\ldots,k}$ is a locally defined ONB for $(T_z\Sigma)^\perp$. Especially

$$\underline{H}(y) \in (T_y\Sigma)^\perp.$$

4. Suppose that $k = 1$ and that $\mathcal{N}$ is a field of unit vectors $\Sigma \to \mathbb{R}^{m+1}$ s.t. $\mathcal{N}(y) \in (T_y\Sigma)^\perp$. Then $\mathcal{N}$ orients Σ and we deduce

$$\underline{H} = (-\operatorname{div}_\Sigma \mathcal{N})\mathcal{N} =: H \cdot \mathcal{N}.$$

The real valued function H is the so called *scalar mean curvature* of Σ. Obviously H depends on the orientation $\mathcal{N}$ of Σ, i.e. H changes sign if we orient Σ by $-\mathcal{N}$. But the above formula for $\underline{H}$ shows once more that the vectorial geometric mean curvature is independent of the possibility of orienting Σ. □

We close this survey with the so-called

Divergence Theorem: *Suppose that $\Sigma \subset \mathbb{R}^{m+k}$ is a m-dimensional manifold perhaps with boundary. Then for any vector field $X \in C_0^1(\mathbb{R}^{m+k}, \mathbb{R}^{m+k})$*

$$\int_\Sigma \operatorname{div}_\Sigma X \, d\mathcal{H}^m = -\int_\Sigma X \cdot \underline{H} \, d\mathcal{H}^m - \int_{\partial\Sigma} X \cdot \eta \, d\mathcal{H}^{m-1}.$$

Here η is a vectorfield defined on $\partial\Sigma$ s.t. $\eta(x) \in (T_\eta\partial\Sigma)^\perp \subset T_\eta\Sigma$ and pointing into Σ. $\mathcal{H}^\ell$ denotes ℓ-dimensional Hausdorff measure on $\mathbb{R}^{m+k}$. If X has support disjoint to $\partial\Sigma$ then no boundary term occurs, i.e.

2.1 The Mean Curvature Problem

$$\int_\Sigma (div_\Sigma X + X \cdot \underline{H}) d\mathcal{H}^m = 0 \qquad (2.1.1)$$

and if in addition X is tangential along Σ then the formula reduces to

$$\int_\Sigma div_\Sigma X \cdot d\mathcal{H}^m = 0.$$

□

Let us now return to the MCP: We have to discuss the precise meaning of prescribing "the mean curvature" of Σ. The first idea could be the following:

> Given a vectorfunction $L : \mathbb{R}^{m+k} \to \mathbb{R}^{m+k}$, find a m–manifold $\Sigma \subset \mathbb{R}^{m+k}$ with $\partial \Sigma = \Gamma$ and the additional property $\int_\Sigma (div_\Sigma X + L \cdot X) d\mathcal{H}^m = 0$ for all $X \in C_0^1(\mathbb{R}^{m+k}, \mathbb{R}^{m+k})$, spt $X \cap \Gamma = \emptyset$.

If the above problem admits a solution Σ then clearly from (2.1.1) we would have $\underline{H}(y) = L(y)$ for all points $y \in \Sigma$ so that the geometric mean curvature vector $\underline{H}$ of our solution Σ is given by the prescribed field L. *But the following simple example shows that in general it does not make sense to "prescribe mean curvature" in terms of a vectorfield depending on the base point only.*

Take $m = 2$, $k = 1$ and assume that the boundary Γ is contained in the plane $\mathbb{R}^2 \times \{0\} \subset \mathbb{R}^3$. Suppose further that we are given the field $L \equiv (0,0,1)$ and that $\Sigma \subset \mathbb{R}^3$ is a solution in the sense described above. From

$$\underline{H}(y) = L(y) = (0, 0, 1),$$

$$\underline{H}(y) \in T_y \Sigma^\perp$$

we then deduce $T_y \Sigma = \mathbb{R}^2 \times \{0\}$, hence Σ is flat and $\underline{H} \equiv 0$.

A more promising approach results from the idea to consider "mean curvature functions" depending not only on the base points but also on the tangent m–planes, i.e. we consider functions

$$\Omega : \mathbb{R}^{m+k} \times G_m(m+k) \to \mathbb{R}^{m+k}$$

with the additional property

$$\Omega(y, E) \in E^\perp \qquad (2.1.2)$$

for all m–planes in the Grassmannian bundle $G_m(m+k)$ (= set of all unoriented m-dimensional subspaces of $\mathbb{R}^{m+k}$). The requirement (2.1.2) is natural in view of the geometric property $\underline{H}(y) \in (T_y \Sigma)^\perp$ of the mean curvature vector. This leads to the following

(Preliminary) Definition: Suppose that $\Omega : \mathbb{R}^{m+k} \times G_m(m+k) \to \mathbb{R}^{m+k}$ satisfies (2.1.2). Σ has prescribed mean curvature function Ω iff $\underline{H}(y) = \Omega(y, T_y\Sigma)$ holds for all $y \in \Sigma$.

In view of the divergence theorem this is equivalent to the equation

$$\int_\Sigma \left(div_\Sigma X + X \cdot \Omega(\cdot, T\Sigma) \right) d\mathcal{H}^m = 0 \qquad (2.1.3)$$

for all $X \in C_0^1(\mathbb{R}^{m+k}, \mathbb{R}^{m+k})$, spt $X \cap \Gamma = \emptyset$. In this generality the mean curvature problem is still unsolved, we have to restrict ourselves to a special class of curvature functions Ω with "nice" dependence on the tangent space argument. In order to get an impression of what the right condition on Ω is we look at the classical case $m = 2$, $k = 1$ for a closed Jordan curve $\Gamma \subset \mathbb{R}^3$ and assume that

$$u : D = \{z \in \mathbb{C} : |z| < 1\} \to \mathbb{R}^3$$

is a smooth mapping satisfying

$$\partial_1 u \cdot \partial_2 u = 0 = |\partial_1 u| - |\partial_2 u|^2 \quad \text{(conformality)} \qquad (2.1.4)$$

$$\partial_1 u \times \partial_2 u \neq 0. \qquad (2.1.5)$$

Then $\Sigma := u(D)$ is a surface in $\mathbb{R}^3$ satisfying

$$\underline{H}_\Sigma(P) = \frac{\Delta u(z)}{|\partial_1 u \times \partial_2 u|(z)}$$

at any point $P = u(z) \in \Sigma$. We equip Σ with its natural orientation

$$\mathcal{N}(P) = \frac{\partial_1 u(z) \times \partial_2 u(z)}{|\partial_1 u(z) \times \partial_2 u(z)|}$$

and observe

$$\underline{H}_\Sigma(P) = H(P) \cdot \mathcal{N}(P)$$

for a scalar function H. We may rewrite this identity as

$$\Delta u = H(u) \, \partial_1 u \times \partial_2 u \quad \text{on } D. \qquad (2.1.6)$$

Conversely one starts from a given Jordan curve Γ and a scalar function $h : \mathbb{R}^3 \to \mathbb{R}$. Then the classical problem of finding a surface of disc type with boundary curve Γ and curvature h is to construct a function u s.t. $u(\partial D) = \Gamma$, that (2.1.4) and (2.1.5) hold and in addition

$$\Delta u(x) = h(u(x)) \, \partial_1 u(x) \times \partial_2 u(x), \quad x \in D. \qquad (2.1.7)$$

2.1 The Mean Curvature Problem

This problem has been successfully attacked by various prominent authors about twenty years ago, we mention the contributions of Heinz [54], Hildebrandt [57] – [59], Gulliver and Spruck [51], Wente [81] and Steffen [76], who gave various conditions on the size of h under which the problem admits a solution $\Sigma = u(D)$. In this case

$$\underline{\underline{H}}_\Sigma(P) = h(P) \cdot \mathcal{N}(P) \tag{2.1.8}$$

and this formula shows that it is only possible to prescribe the length $|\underline{\underline{H}}_\Sigma|$ $(= |h|)$ of the geometric mean curvature vector but not its direction: through the dependence

$$\mathcal{N}(P) = \frac{\partial_1 u(z) \times \partial_2 u(z)}{|\partial_1 u(z) \times \partial_2 u(z)|}$$

the direction $\mathcal{N}$ is a function of the unknown surface $u(D) = \Sigma$.

Define $\Lambda^2 \mathbb{R}^3$ as the space of all alternating 2–linear mappings $\varphi : \mathbb{R}^3 \times \mathbb{R}^3 \to \mathbb{R}$, i.e. φ is linear in each argument and $\varphi(\tau, \eta) = -\varphi(\eta, \tau)$. By definition $\Lambda_2 \mathbb{R}^3$ is the dual space of $\Lambda^2 \mathbb{R}^3$ and generated by the exterior products $\tau \wedge \eta$, $\tau, \eta \in \mathbb{R}^3$. Notice that we have a natural isomorphism $\Lambda_2 \mathbb{R}^3 \ni \tau \wedge \eta \mapsto \tau \times \eta \in \mathbb{R}^3$ with the help of the vector product.

Let $\Omega : \mathbb{R}^3 \times \Lambda_2 \mathbb{R}^3 \to \mathbb{R}^3$,

$$\Omega(P, \tau \wedge \eta) := h(P)\, \tau \times \eta \,.$$

Then (2.1.8) can be rewritten as

$$\underline{\underline{H}}_\Sigma(P) = \Omega(P, \tau_1 \wedge \tau_2)$$

where τ_1, τ_2 is a positive ONB in $T_p\Sigma$. The function $\Omega(P, \cdot)$ defined above clearly extends to a linear function on the whole of $\Lambda_2 \mathbb{R}^3$ and this leads us to consider curvature functions

$$\Omega : \mathbb{R}^{m+k} \times \Lambda_m \mathbb{R}^{m+k} \to \mathbb{R}^{m+k}$$

which are *linear* in the second argument. Unfortunately this forces us to work in the class of oriented manifolds (although the mean curvature vector $\underline{\underline{H}}_\Sigma$ making sense for unoriented m–manifolds) but we preferred this restricted case for several reasons:

- We did not succeed in making any substantial contribution concerning the case of general curvature forms $\mathbb{R}^{m+k} \times G_m(m+k) \to \mathbb{R}^{m+k}$.

- As we shall see below the discussion of linear curvature forms in a class of orientable objects has a natural variational formulation in the context of Geometric Measure Theory.

Let us now give precise definitions of our setting. We use "multi-index" notation in which $\alpha = (\alpha_1, \ldots, \alpha_m)$ and $I_m := \{\alpha : 1 \leq \alpha_1 < \alpha_2 < \ldots < \alpha_m \leq m+k\}$. An m–*covector* is an m–linear alternating map

$$\omega : \mathbb{R}^{m+k} \times \ldots \times \mathbb{R}^{m+k} \to \mathbb{R}.$$

The set $\Lambda^m \mathbb{R}^{m+k}$ of all m–covectors forms a vectorspace of dimension $\binom{m+k}{m}$ with basis

$$\{dx^\alpha = dx^{\alpha_1} \wedge \ldots \wedge dx^{\alpha_m} : \alpha \in I_m\}$$

and dx^i is the 1–covector given by

$$dx^i(v) := v \cdot e_i.$$

The dual space $\Lambda_m \mathbb{R}^{m+k}$ consisting of all m–vectors has the dual basis $\{e_\alpha = e_{\alpha_1} \wedge \ldots \wedge e_{\alpha_m} : \alpha \in I_m\}$. The elements $\tau_1 \wedge \ldots \wedge \tau_m$ of $\Lambda_m \mathbb{R}^{m+k}$ with $\tau_i \in \mathbb{R}^{m+k}$ are called simple. One can show

$$\tau_1 \wedge \ldots \wedge \tau_m \neq 0 \iff \tau_i, \quad i = 1, \ldots, m, \text{ are linearly independent.}$$

The simple m–vectors $\xi = \tau_1 \wedge \ldots \wedge \tau_m \neq 0$ are in unique correspondence with the m–dimensional subspaces of $\mathbb{R}^{m+k}$ by defining $E_\xi = \{v \in \mathbb{R}^{m+k} : v \wedge \xi = 0\}$. Moreover, $E_\xi = E_{\xi'}$ implies $\xi' = c \cdot \xi$ for some $c \in \mathbb{R}$, hence $\xi' = \pm \xi$ if we consider vectors of length one. Thus we may identify

$$G_m^{or}(m+k) = \text{oriented } m\text{–planes in } \mathbb{R}^{m+k}$$

and

$$\{\xi \in \Lambda_m \mathbb{R}^{m+k} : \xi \text{ simple}, |\xi| = 1\}.$$

From a geometrical point of view any $\xi = \tau_1 \wedge \ldots \wedge \tau_m \neq 0$ can be interpreted as a "vector" perpendicular to the plane E_ξ. In case $m = 2$, $k = 1$ this interpretation can be made concrete by identifying $\tau_1 \wedge \tau_2$ and $\tau_1 \times \tau_2 \in \mathbb{R}^3$.

Let us also remark that an m–covector may be thought of as acting on either an m–tuple $(v_1, \ldots, v_m)$ linearly and alternating or in a linear way on a single m–vector.

Definition (oriented mean curvature forms:) An oriented mean curvature form is a mapping

$$\Omega : \mathbb{R}^{m+k} \times \Lambda_m \mathbb{R}^{m+k} \to \mathbb{R}^{m+k}$$

such that

(i) $\Omega(P, \cdot)$ is a linear mapping $\Lambda_m \mathbb{R}^{m+k} \to \mathbb{R}^{m+k}$ for all $P \in \mathbb{R}^{m+k}$.

2.1 The Mean Curvature Problem

(ii) $\Omega(P, \tau_1 \wedge \ldots \wedge \tau_m) \in Span\,[\tau_1, \ldots, \tau_m]^\perp$ for all $P \in \mathbb{R}^{m+k}$, $\tau_1, \ldots, \tau_m \in \mathbb{R}^{m+k}$.

Remarks:

1. $\Omega(P, \cdot)$ can be viewed as an m–linear alternating mapping $\mathbb{R}^{m+k} \times \ldots \times \mathbb{R}^{m+k} \to \mathbb{R}^{m+k}$ or as an m–covector with values in $\mathbb{R}^{m+k}$.

2. Another interpretation of $\Omega(P, \cdot)$ is that of a function
$$G_m^{or}(\mathbb{R}^{m+k}) \to \mathbb{R}^{m+k}$$
with linear extension to the whole space $\Lambda_m \mathbb{R}^{m+k}$.

3. It is easy to construct mean curvature forms: take any $\omega \in \Lambda^{m+1} \mathbb{R}^{m+k}$ and consider
$$\varphi : \mathbb{R}^{m+k} \ni v \mapsto \omega(v \wedge \tau_1 \wedge \ldots \wedge \tau_m) \in \mathbb{R}$$
for $\tau_1, \ldots, \tau_m \in \mathbb{R}^{m+k}$ fixed. Clearly there exist $\Omega(\tau_1 \wedge \ldots \wedge \tau_m) \in \mathbb{R}^{m+k}$ such that
$$\varphi(v) = v \cdot \Omega(\tau_1 \wedge \ldots \wedge \tau_m),$$
and Ω satisfies i), ii) from above. In this special case the oriented mean curvature form Ω is independent of the base point P but by applying the construction to ω's depending also on P we cover the general case.

Definition: If $\Gamma \subset \mathbb{R}^{m+k}$ is an oriented $(m-1)$-dimensional manifold and if an oriented mean curvature form Ω is given then an *oriented m–manifold Σ has boundary Γ and prescribed mean curvature form Ω* iff $\partial \Sigma = \Gamma$ and

$$\int_\Sigma \left(div_\Sigma X + X \cdot \Omega(\cdot, \vec{\Sigma}) \right) d\mathcal{H}^m = 0 \qquad (2.1.9)$$

holds for all $X \in C_0^1(\mathbb{R}^{m+k}, \mathbb{R}^{m+k})$, $spt\,X \cap \Gamma = \emptyset$. Here $\vec{\Sigma}(P) = \tau_1(P) \wedge \ldots \wedge \tau_m(P)$ for some positive ONB $\{\tau_1(P), \ldots, \tau_m(P)\}$ in $T_P \Sigma$.

If the M.C.P. has a solution in the above sense then
$$\underline{\underline{H}}_\Sigma = \Omega(\cdot, \vec{\Sigma})$$
holds for the geometric mean curvature vector.

Remarks:

1. If Σ has prescribed mean curvature form Ω in the sense of equation (2.1.9) then the manifold Σ^- obtained by changing the orientation has mean curvature form $-\Omega$ since $\Omega(\cdot, \vec{\Sigma^-}) = -\Omega(\cdot, \vec{\Sigma})$. Hence we may rewrite (2.1.9) as

$$\int_\Sigma \left(div_\Sigma X + (-\Omega)(\cdot, \vec{\Sigma^-}) \right) d\mathcal{H}^m = 0$$

which shows that Σ^- is a solution to the M.C.P. with curvature form $-\Omega$. But clearly $\underline{\underline{H}}_\Sigma = \underline{\underline{H}}_{\Sigma^-}$.

2. As remarked earlier one should try to study unoriented mean curvature forms, e.g. functions $\overline{\Omega} : \mathbb{R}^{m+k} \times \Lambda_m \mathbb{R}^{m+k} \to \mathbb{R}^{m+k}$ with the following properties:

   (i) $\overline{\Omega}(P, c \cdot \xi) = |c| \cdot \overline{\Omega}(P, \xi), \quad c \in \mathbb{R}$
   $$\left(\Rightarrow \overline{\Omega}(P, -\xi) = \overline{\Omega}(P, \xi) \right)$$

   (ii) $\overline{\Omega}(P, \tau_1 \wedge \ldots \wedge \tau_m) \in Span[\tau_1, \ldots, \tau_m]^\perp$

By i) $\overline{\Omega}$ is indpendent of the orientation and in place of (2.1.9) we would have to solve

$$\int_\Sigma \left(div_\Sigma X + X \cdot \overline{\Omega}(\cdot, T\Sigma) \right) d\mathcal{H}^m = 0$$

for some (unoriented) manifold Σ where $T\Sigma$ represents the m-vector $\tau_1 \wedge \ldots \wedge \tau_m$ for an arbitrary choice of ONB $\{\tau_i\}$ in the tangent space. But this more general problem has no variational structure and therefore can not be covered by our methods. Nevertheless it would be very interesting to treat the nonoriented curvature problem for the reason that the mean curvature vector $\underline{\underline{H}}_\Sigma$ makes sense for nonoriented manifolds. On the other hand the classical parametric approach via mappings $D \to \mathbb{R}^3$ always produces orientable surfaces of disc type. Jost proposed (in a problem list on geometric variational problems) the study of the existence of solutions to $\Delta u = H\left(u, \frac{\partial_1 u \times \partial_2 u}{|\partial_1 u \times \partial_2 u|}\right)$ satisfying (2.1.4), (2.1.5) and Plateau's boundary condition for a function $H : \mathbb{R}^3 \times S^2 \to \mathbb{R}^3$ with $H(x, \cdot)$ satisfying $H(x, \eta) = H(x, -\eta)$ which corresponds to our generalisation but up to now there has been no contribution. $\square$

We have now given a precise definition of the M.C.P. in the setting of oriented smooth manifolds. From the classical theory it is well known that

2.1 The Mean Curvature Problem

one can not expect solvability of the problem for arbitrary large curvatures: Heinz [55] showed the following. Let $\Gamma = \partial D$ denote the unit circle in $\mathbb{R}^2$. Then, if $h \in \mathbb{R}$ such that $|h| > 2$ the parametric M.C.P.

$$\begin{cases} \Delta u = h \cdot \partial_1 u \times \partial_2 u \text{ on } D, \quad u(\partial D) = \Gamma \\ |\partial_1 u|^2 - |\partial_2 u|^2 = 0 = \partial_1 u \cdot \partial_2 u \end{cases}$$

has *no* solution.

Consider now a general $(m-1)$–dimensional oriented manifold and an oriented mean curvature form Ω whre for simplicity Ω is assumed to be independent of the base point which corresponds to the case $h \in \mathbb{R}$ in the parametric setting. Let Σ denote a solution of the *M.C.P.* The divergence theorem implies

$$\int_\Sigma \left(div_\Sigma X + X \cdot \Omega(\vec{\Sigma}) \right) d\mathcal{H}^m = \int_\Gamma X \cdot \eta \, d\mathcal{H}^{m-1}$$

for $X \in C_0^1(\mathbb{R}^{m+k}, \mathbb{R}^{m+k})$. Assuming that the solution Σ stays in a compact region the above formula extends to X with arbitrary support, especially we may choose $X \equiv e \in \mathbb{R}^{m+k}$ and get

$$\int_\Sigma e \cdot \Omega(\vec{\Sigma}) \, d\mathcal{H}^m = \int_\Gamma e \cdot \eta \, d\mathcal{H}^{m-1}$$

thus

$$\left| \int_\Sigma e \cdot \Omega(\vec{\Sigma}) \, d\mathcal{H}^m \right| \leq \mathcal{H}^{m-1}(\Gamma)$$

since $|\eta| = 1$ and we may also take $|e| = 1$.

Next let S denote a *minimal* submanifold for the boundary Γ, i.e. a smooth oriented m–manifold with $\partial S = \Gamma$ and

$$\mathcal{H}^m(S) \leq \mathcal{H}^m(S') \; \forall S', \quad \partial S' = \Gamma.$$

After a possible change of orientation $\Sigma \cup S$ is a closed oriented manifold $\mathcal{M}$, hence

$$\int_\mathcal{M} \omega(\vec{\mathcal{M}}) \cdot d\mathcal{H}^m = 0$$

for all m–forms ω with $d\omega = 0$.

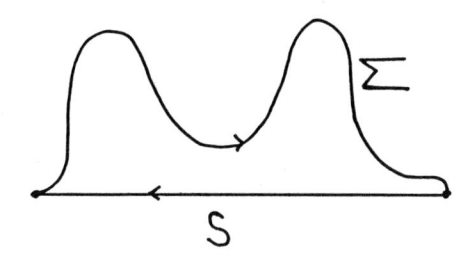

Here we have
$$\omega_x(\tau_1,\ldots,\tau_m) := e \cdot \Omega(\tau_1 \wedge \ldots \wedge \tau_m)$$
and ω is independent of x so that $d\omega = 0$. This implies
$$\left| \int_\Sigma e \cdot \Omega(\vec{\Sigma}) d\mathcal{H}^m \right| = \left| \int_S e \cdot \Omega(\vec{S}) d\mathcal{H}^m \right|$$
and we arrive at the following

necessary condition for the solvability of the M.C.P. : $\left| \int_S e \cdot \Omega(\vec{S}) d\mathcal{H}^m \right| \leq \mathcal{H}^{m-1}(\Gamma) \quad \forall |e| = 1$.
(2.1.10)

Clearly our arguments are somewhat unprecise since for $k > 1$ it is by no means clear (and in general false!) that there are *smooth* minimal objects S for the boundary Γ. (This is true only for $k = 1$ and $m \leq 6$.) But condition (2.1.10) can be rewritten in terms of generalized manifolds ("currents") which at this stage has no advantage since we will end up with the same results.

For example let $\Gamma = \partial B^m$ with
$$B^m = \{(x,0) \in \mathbb{R}^{m+k} : x \in \mathbb{R}^m,\ |x| < 1\}.$$

Then the "minimal surface" S is also contained in $\mathbb{R}^m \times \{0\}$ and clearly is given by B^m. If we further assume $\underline{k=1}$ then $\vec{S} = e^{m+1}$ so that
$$\Omega(\vec{S}) \equiv \lambda e^{m+1}$$
for some $\lambda \in \mathbb{R}$. Let $e = e^{m+1}$. Then (2.1.10) reduces to $|\lambda| \cdot \mathcal{H}^m(B^m) \leq \mathcal{H}^{m-1}(\partial B^m)$. Since $\mathcal{H}^m(B^m) = m \cdot \mathcal{H}^{m-1}(\partial B^m)$ we arrive at Heinz's condition
$$|\lambda| \leq m,$$
i.e. for $|\lambda| > m$ we can not expect solvability of the M.C.P. for the special configuration.

Let us return to the discussion of (2.1.10) in the general case: Obviously (2.1.10) holds if we require
$$|\Omega| \leq \mathcal{H}^{m-1}(\Gamma)/\mathcal{H}^m(S). \tag{2.1.11}$$

On the other hand we may argue as follows: if $\Gamma \subset \overline{B}_R^{m+k}(x)$ for some ball then also $S \subset \overline{B}_R^{m+k}(x)$ by the convex hull property of minimal surfaces. From Almgren's Optimal Isoperimetric Inequality [3] we infer

2.1 The Mean Curvature Problem

$$\mathcal{H}^m(S) \leq c \cdot \mathcal{H}^{m-1}(\Gamma)^{\frac{m}{m-1}},$$

$$c := m^{-1-\frac{1}{m-1}} \cdot \alpha_m^{-1/m-1}, \quad \alpha_m = \mathcal{H}^m(B_1^m(0))$$

(2.1.12)

with " $=$ " iff S is an m–dimensional ball.

Applying (2.1.12) we see

$$\begin{aligned}
\mathcal{H}^{m-1}(\Gamma)/\mathcal{H}^m(S) &\geq \left(\frac{1}{c} \cdot \mathcal{H}^m(S)\right)^{\frac{m-1}{m}} \cdot \mathcal{H}^m(S)^{-1} \\
&= \left(\frac{1}{c}\right)^{1-\frac{1}{m}} \mathcal{H}^m(S)^{-1/m} \\
&= m \cdot \alpha_m^{1/m} \mathcal{H}^m(S)^{-1/m}
\end{aligned}$$

and (2.1.11) is a consequence of

$$\boxed{|\Omega| \leq m \cdot \alpha_m^{1/m} \cdot \mathcal{H}^m(S)^{-1/m}.} \tag{2.1.13}$$

Recalling $S \subset \overline{B}_R^{m+k}(x)$ we deduce

$$\mathcal{H}^m(S) \leq \alpha_m \cdot R^m$$

and (2.1.13) follows from the requirement

$$|\Omega| \leq m \cdot R^{-1}. \tag{2.1.14}$$

Proposition: *For constant mean curvature forms Ω and boundaries $\Gamma \subset \overline{B}_R^{m+k}(x)$ we have (2.1.10)*

$$\left|\int_S e \cdot \Omega(\vec{S}) d\mathcal{H}^m\right| \leq \mathcal{H}^{m-1}(\Gamma), \quad \forall \, |e| = 1$$

as a necessary condition for solvability of the M.C.P. (2.1.10) holds under the assumption (2.1.13)

$$|\Omega| \leq m \cdot \alpha_m^{1/m} \mathcal{H}^m(S)^{-1/m}$$

and (2.1.13) is true in case (2.1.14)

$$|\Omega| \leq m \cdot R^{-1}.$$

In case $k = 1$, $\Gamma = \partial B_1^m(0)$ all these conditions are equivalent. □

For long but small boundary configuration Γ the radius of the ball $B_R^{m+k}(x)$ containing Γ has to be chosen rather large in comparison to the area $\mathcal{H}^m(S)$ of the correponding minimal surface. So for boundaries which are far from being spherical (2.1.13) allows larger values of the mean curvature form Ω. It is therefore desirable to prove existence of solutions under conditions of the form (2.1.13), i.e. to show that (2.1.13) is sufficient for the existence of solutions to the M.C.P.

As a consequence of our general theorems in section 2.3 we obtain the following

Theorem: *Assume $k = 1$, $m \leq 6$ and let Γ denote a smooth oriented $(m-1)$-manifold without boundary. Assume that $\Omega : \mathbb{R}^{m+1} \times \Lambda_m \mathbb{R}^{m+1} \to \mathbb{R}^{m+1}$ is a mean curvature form such that*

$$|\Omega| < m \cdot \left(\alpha_{m+1}^{1/m} \cdot 2^{-1/m} \right) \cdot \mathcal{H}^m(S)^{-1/m} \qquad (2.1.15)$$

holds when S minimizes area. Then the M.C.P. admits at least one solution Σ.

Remarks:

1. We have $|\Omega| := \sup_P |\Omega_P|$ and

$$|\Omega_p| := \sup_{\substack{\tau_i \in \mathbb{R}^{m+1} \\ \tau_i \cdot \tau_j = \delta_{ij}}} |\Omega_p(\tau_1 \wedge \ldots \wedge \tau_m)|.$$

2. Define

$$\Omega_0(\tau_1 \wedge \ldots \wedge \tau_m) := \sum_{m+1}^{i=1} (-1)^{i+1} \det\left(\widehat{\tau_1^i}, \ldots, \widehat{\tau_m^i} \right) e_i,$$

$$\widehat{\tau_k^i} = \sum_{j=1, j \neq i}^{m+1} \tau_k^j e_j, \quad k = 1, \ldots, m, \quad i = 1, \ldots, m+1,$$

where $\tau_k = \left(\tau_k^i \right)_{1 \leq i \leq m+1}$, $k = 1, \ldots, m$. Then $\Omega_0(\tau_1 \wedge \ldots \wedge \tau_m) \in Span[\tau_1, \ldots, \tau_m]^\perp$ and for dimensional reasons we have at each point $P \in \mathbb{R}^{m+1}$

$$\Omega_P = H(P) \cdot \Omega_0$$

with a scalar function $H : \mathbb{R}^{m+1} \to \mathbb{R}$. Thus in the codimension 1 case any mean curvature form is uniquely determined by a function $H : \mathbb{R}^{m+1} \to \mathbb{R}$ and (2.1.15) can be seen as a bound on $\sup_{\mathbb{R}^{m+1}} |H|$.

Acutally it is possible to replace the sup–norm of H by the L^{m+1}–norm of H which is more appropriate for applications.

3. (2.1.15) is slightly stronger than the "necessary" condition (2.1.13) since $\sqrt[m]{\frac{\alpha_{m+1}}{2}} < \sqrt[m]{\alpha_m}$. For example if $m = 2$ and $\Gamma = S^1 = \{(x,y,0) \in \mathbb{R}^3 : x^2 + y^2 = 1\}$ then (2.1.13) reads as $|\Omega| \leq 2$ and (2.1.15) reduces to $|\Omega| < 2 \cdot \sqrt{\frac{2}{3}\pi}/\sqrt{\pi} = 2 \cdot \sqrt{\frac{2}{3}} \approx 1.633$.

Thus we have existence for the M.C.P. under the curvature assumption $|H| < 1.633$ and existence can in general not be expected for $|H| > 2$. For $H \in \mathbb{R}$ and the corresponding parametric problem where Γ denotes an arbitrary Jordan curve in $B_1^3(0)$ Hildebrandt proved existence in case $|H| \leq 2$. To my opinion it is a very hard problem to obtain existence for the general M.C.P. under the assumption (2.1.13) $|\Omega| \leq m \cdot \alpha_m^{1/m} \mathcal{H}^m(S)^{-1/m}$.

2.2 Some Facts from Geometric Measure Theory

The standard references are Federer's book [23] or more updated the recent monography of L. Simon [74].

Suppose $m \geq 2$, $k = 1$ and let Γ denote a smooth oriented $(m-1)$-manifold with $\partial \Gamma = \emptyset$, w.l.o.g. we may assume $\Gamma = \partial \Sigma_0$ for an m-manifold Σ_0. For a given mean curvature form $\Omega : \mathbb{R}^{m+k} \times \Lambda_m \mathbb{R}^{m+k} \to \mathbb{R}^{m+k}$ we want to solve the

$$
\text{(M.C.P.)} \begin{cases} \text{find } \Sigma \text{ such that } \partial \Sigma = \partial \Sigma_0 \text{ and} \\ \int_\Sigma (\text{div}_\Sigma X + X \cdot \Omega(\vec{\Sigma})) \, d\mathcal{H}^m = 0 \\ \text{for all } X \in C_0^1(\mathbb{R}^{m+k}, \mathbb{R}^{m+k}), \text{spt } X \cap \Gamma = \emptyset. \end{cases}
$$

We impose the following condition on Ω: Let

$$\omega : \mathbb{R}^{m+k} \times \Lambda_{m+1} \mathbb{R}^{m+k} \to \mathbb{R},$$

$$\omega_p(\tau_1 \wedge \ldots \wedge \tau_{m+1}) := \tau_1 \cdot \Omega_p(\tau_2 \wedge \ldots \wedge \tau_{m+1})$$

denote the scalar mean curvature form associated to Ω. We assume

$$d\omega = 0. \tag{2.2.1}$$

In this case $\int_Q \omega(\vec{Q}) d\mathcal{H}^{m+1} = 0$ by Stoke's Theorem for any closed $(m+1)$-manifold. Note that (2.2.1) trivially holds in case $k = 1$ and for general $k \geq 2$ if Ω is constant, i.e. independent of the base point.

Our approach to the M.C.P. uses a variational argument: consider any Σ such that $\partial\Sigma = \partial\Sigma_0$. Then $\Sigma \cup \Sigma_0$ (equipped with suitable orientation) is a closed m–manifold and we select a $(m+1)$–manifold Q such that $\partial Q = \Sigma \cup \Sigma_0$ (Note that for $k \geq 2$ Q is not unique). We define

$$\mathcal{F}(\Sigma) := \mathcal{H}^m(\Sigma) + \int_Q \omega(\vec{Q})\, d\mathcal{H}^{m+1}$$

and note $\int_Q \omega(\vec{Q})d\mathcal{H}^{m+1} = \int_{Q'} \omega(\vec{Q'})d\mathcal{H}^{m+1}$ as long as $\partial Q = \partial Q'$.

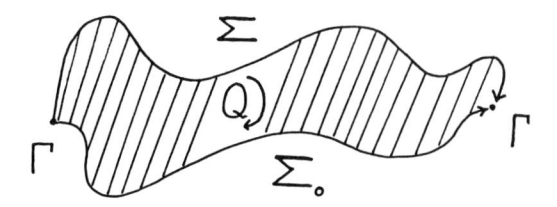

For $X \in C_0^1(\mathbb{R}^{m+k}, \mathbb{R}^{m+k})$ with $\operatorname{spt} X \cap \Gamma = \emptyset$ we let

$$\Phi_t(x) := x + t \cdot X(x)$$

and consider the deformed manifold $\Sigma_t := \Phi_t(\Sigma)$. It can be shown that

$$\frac{d}{dt/0}\mathcal{F}(\Sigma_t) = \int_\Sigma (\operatorname{div}_\Sigma X + X \cdot \Omega(\vec{\Sigma}))\, d\mathcal{H}^m$$

which suggests to obtain solutions of the M.C.P. by considering

$$\mathcal{F} \to \operatorname{Min}$$

in the class of all m–manifolds Σ with boundary Γ. Here we encounter the same difficulties as in parametric variational problems which we studied in Chapter 1: it is impossible to study the problem by direct methods in spaces of smooth objects. This is not only a technical problem since even in the case $\Omega = 0$ area minimizing objects may behave irregular if $m \geq 7$ or $k \geq 2$. So we have to single out classes of generalized m–dimensional objects for which we can define boundaries and for which the Mean Curvature Equation makes sense. We introduce some

Notations form Geometric Measure Theory:
For U open in $\mathbb{R}^{m+k}$ we let

$$\mathcal{D}^\ell(U) = \text{space of all smooth real } \ell\text{–forms with compact support in } U.$$

2.2 Some Facts from Geometric Measure Theory

We write

$$\omega = \sum_{\alpha \in I_\ell} \omega^\alpha \, dx^\alpha,$$

$$I_\ell = \{\alpha = (\alpha_1, \ldots, \alpha_\ell) : 1 \leq \alpha_1 < \alpha_2 < \ldots < \alpha_\ell \leq m+k\},$$

$$dx^\alpha = dx^{\alpha_1} \wedge \ldots \wedge dx^{\alpha_\ell},$$

$$|\omega| := \sup_{x \in U} \left(\sum_{\alpha \in I_\alpha} \omega^\alpha(x)^2 \right)^{1/2}$$

(Euclidean norm).

For any $x \in U$ ω_x acts as a linear map $\Lambda_\ell \mathbb{R}^{m+k} \to \mathbb{R}$. Convergence $\omega_i \xrightarrow[i \to \infty]{} \omega$ in $\mathcal{D}^\ell(U)$ is defined as follows: there exists a compact set $K \subset U$ such that

$$\operatorname{spt} \omega_i \cup \operatorname{spt} \omega \subset K$$

and

$$\omega_i^\alpha \xrightarrow[i \to \infty]{} \omega^\alpha$$

uniformly with all partial derivatives, $\alpha \in I_\ell$.

Definition (ℓ–currents on U): An ℓ–current T is a continuous linear map $\mathcal{D}^\ell(U) \to \mathbb{R}$ acting on smooth ℓ–forms with compact support in U.

Continuity means

$$\omega_i \xrightarrow[i \to \infty]{} \omega \text{ in } \mathcal{D}^\ell(U) \implies \lim_{i \to \infty} T(\omega_i) = T(\omega).$$

By definition the ℓ–currents are just the distributions on the space $\mathcal{D}^\ell(U)$. The space of all ℓ–currents is denoted by $\mathcal{D}_\ell(U)$.

Consider an oriented smooth ℓ-dimensional submanifold of $\mathbb{R}^{m+k}$ with boundary $\partial\Sigma$ such that $\Sigma \cup \partial\Sigma \subset\subset U$. We may integrate ℓ–forms φ over Σ by the way obtaining the special ℓ–current

$$[\![\Sigma]\!] \in \mathcal{D}_\ell(U), \quad [\![\Sigma]\!](\varphi) := \int_\Sigma \varphi(\vec{\Sigma}) \, d\mathcal{H}^\ell.$$

("integration over Σ")

where

$$\vec{\Sigma}(P) = \tau_1(P) \wedge \ldots \wedge \tau_\ell(P) \in \Lambda_\ell \mathbb{R}^{m+k}$$

for some positive ONB $\{\tau_i(P)\}$ in $T_p\Sigma$. One easily proves

$$\mathcal{H}^\ell(\Sigma) = \sup_{\substack{\omega \in \mathcal{D}^\ell(U), \\ |\omega| \leq 1}} [\![\Sigma]\!](\omega)$$

and Stoke's Theorem implies

$$[\![\partial \Sigma]\!](\psi) = [\![\Sigma]\!](d\psi), \quad \psi \in \mathcal{D}^{\ell-1}(U).$$

This gives rise to the following:

Definition: For $T \in \mathcal{D}_\ell(U)$ we define
the mass (or area): $\underline{M}(T) := \sup_{\substack{\omega \in \mathcal{D}^\ell(U), \\ |\omega| \leq 1}} T(\omega)$,
the boundary $\partial T \in \mathcal{D}_{\ell-1}(U) : \partial T(\varphi) := T(d\varphi), \varphi \in \mathcal{D}^{\ell-1}(U)$,
the support: $\operatorname{spt}(T) := U - \bigcup \{V \text{ open in } U : T(\omega) = 0$
for all $\omega \in \mathcal{D}^\ell(V)\}$.

Smooth oriented manifolds induce currents but the converse is in general not true. So one singles out subclasses of $\mathcal{D}_\ell(U)$ with additional structure.

If for $T \in \mathcal{D}_\ell(U)$ $\underline{M}(T) < \infty$ then by Riesz-Representation there is a Radon measure μ_T on U and a μ_T-measurable function

$$\vec{T} : U \to \Lambda_\ell \mathbb{R}^{m+k}$$

such that $|\vec{T}(x)| = 1$ μ_T – a.e. and

$$T(\omega) = \int \langle \omega, \vec{T} \rangle \, d\mu_T, \quad \omega \in \mathcal{D}^\ell(U).$$

Thus the currents T of finite mass are representable by integration.

Lemma 2.2.1 *Suppose that $\{T_i\}$ is a sequence in $\mathcal{D}_\ell(U)$.*

(i)   *If there exists a current $T \in \mathcal{D}_\ell(U)$ such that $\lim_{i \to \infty} T_i(\varphi) = T(\varphi)$ for all $\varphi \in \mathcal{D}^\ell(U)$ (we write $T_i \to T$ in this case) then*

$$\underline{M}(T) \leq \liminf_{i \to \infty} \underline{M}(T_i)$$

("lower semicontinuity of mass w.r.t. current convergence")

(ii) *If $\sup_{i \in \mathbb{N}} \underline{M}(T_i) < \infty$ then there is a subsequence $\{T_i^*\}$ and a current $T \in \mathcal{D}_\ell(U)$ such that $T_i^* \to T^*$ in $\mathcal{D}_\ell(U)$.*

2.2 Some Facts from Geometric Measure Theory

Now, if Γ is an $(m-1)$–dimensional boundary (current $\in \mathcal{D}_{m-1}(U)$, $\partial \Gamma = 0$) and if there exists an m–current of finite mass and boundary Γ then with the help of the above Lemma we can easily produce solutions of

$$\underline{M}(T) \to \min \quad \text{in } \mathcal{D}_m(U), \; \partial T = \Gamma$$

by considering minimizing sequences. The solution would be a current of finite mass but this class is still too big to show smoothness (under restrictions on m and k) of the current solution. Before introducing the "correct" class we briefly discuss the concept of *mapping currents*.

Definition: Suppose that $T \in \mathcal{D}_\ell(U)$ has finite mass and consider a map of class C^1 $f: U \to V$ with range in some open set $V \subset \mathbb{R}^p$ and the property that $f|_{\text{spt } T}$ is *proper*. The push-forward $f_\# T \in \mathcal{D}_\ell(V)$ is the current

$$(f_\# T)(\omega) = \int \langle \omega(f(x)), \Lambda Df(x)\vec{T}(x)\rangle d\mu_T(x), \quad \omega \in \mathcal{D}^\ell(V),$$

where $\Lambda Df(x): \Lambda_\ell \mathbb{R}^{m+k} \to \Lambda_\ell \mathbb{R}^p$ denotes the natural linear map

$$\tau_1 \wedge \ldots \wedge \tau_\ell \mapsto \partial_{\tau_1} f(x) \wedge \ldots \wedge \partial_{\tau_\ell} f(x) \quad (+ \text{ unique linear extension}).$$

One has the mass bound

$$\underline{M}(f_\# T) \leq (\text{Lip}(f))^\ell \cdot \underline{M}(T). \tag{2.2.2}$$

If also $\underline{M}(\partial T) < \infty$ then the push forward $f_\#(\partial T) \in \mathcal{D}_{\ell-1}(V)$ is defined and

$$\partial(f_\# T) = f_\#(\partial T). \tag{2.2.3}$$

It should be noted that (by approximation) the notion $f_\# T$ extends to non smooth Lipschitz $f: U \to V$ with $f|_{\text{spt } T}$ proper. (2.2.2), (2.2.3) continue to hold in this more general setting which is of importance for practical applications.

Definition: A current $T \in \mathcal{D}_\ell(U)$ is said to be an integer multiplicity ℓ–rectifiable current if T can be written as

$$T(\omega) = \int_M \langle \omega, \xi\rangle \Theta \cdot d\mathcal{H}^\ell \quad \forall \omega \in \mathcal{D}^\ell(U)$$

where

M denotes an $\mathbb{H}^\ell$-measurable, locally ℓ-rectifiable subset of U s.t. $\mathcal{H}^\ell(M \cap K) < \infty$ for any compact $K \subset U$,

$\Theta : M \to \mathbb{N}$ is locally $\mathcal{H}^\ell$-integrable

and

$\xi : M \to \Lambda_\ell \mathbb{R}^{n+k}$ is $\mathcal{H}^\ell$-measurable

with the additional property $\xi(x) = \tau_1(x) \wedge \ldots \wedge \tau_\ell(x)$ $\mathcal{H}^\ell$-a.e. on M where $\{\tau_i(x)\}$ is an ONB for $T_x M$.
We write $\mathcal{R}_\ell(U)$ for this class of currents and use the standard notation $T = \underline{\tau}(M, \xi, \Theta)$.

Remarks:

1. Since by definition $T \in \mathcal{R}_\ell(U)$ is of locally finite mass we deduce $\mu_T = \mathcal{H}^\ell \llcorner \Theta$, $\vec{T} = \xi$.

2. Let us recall that an $\mathcal{H}^\ell$-measurable set $M \subset U$ is *locally $\mathcal{H}^\ell$-rectifiable* iff M can be decomposed as

$$M \subset N_0 \cup \bigcup_{j=1}^{\infty} N_j$$

for a set N_0 with $\mathcal{H}^\ell$-measure 0 and ℓ-dimensional smooth submanifolds N_j. So M splits into a set of $\mathcal{H}^\ell$-measure 0 and a countable union of pieces of smooth ℓ-dimensional manifolds. Such a set M has nice tangent properties: it can be shown that for a.a. points $x \in M$ a unique ℓ-dimensional subspace E (called $T_x M$) exists such that

$$\lim_{\lambda \downarrow 0} \int_M f(\frac{1}{\lambda}(z - x)) \, d\mathcal{H}^\ell(z) = \int_{E_x} f \cdot d\mathcal{H}^\ell \quad \forall f \in C_0^0(\mathbb{R}^{n+k}).$$

(For smooth ℓ-manifolds this is just another definition for $T_x M$.)

Since we have tangent spaces *the concept of relative divergence* $\mathrm{div}_M X$, $X \in C_0^1(U, \mathbb{R}^{m+k})$ can be defined just as in section 2.1 and gives rise to a function making sense at $\mathcal{H}^\ell$-almost all points of M:

$$\mathrm{div}_M X(x) := \sum_{i=1}^{\ell} \partial_{\tau_i} X(x) \cdot \tau_i$$

provided $T_x M$ exists; $\{\tau_i\}$ is any ONB in $T_x M$.

2.2 Some Facts from Geometric Measure Theory

3. Any smooth oriented ℓ–manifold $\Sigma \subset U$ induces a current $[\![\Sigma]\!] \in \mathcal{R}_\ell(U)$:

$$[\![\Sigma]\!] = \underline{\tau}(\Sigma, \vec{\Sigma}, 1) .$$

Conversely the elements $T \in \mathcal{R}_\ell(U)$ can be seen as objects which are rather close to the smooth case: T is a singular manifold with oriented tangent spaces $\mathcal{H}^\ell$ – a.e. and the additional property that different pieces of the underlying supporting set M are counted with possibly different weights Θ.

We now state two very important theorems on integer multiplicty currents.

Compactness Theorem (Federer & Fleming, [24]): Let T_j denote a sequence of currents such that

$$T_j \in \mathcal{R}_\ell(U), \quad \partial T_j \in \mathcal{R}_{\ell-1}(U) \quad \text{and}$$

$$\sup_j \{\underline{M}(T_j) + \underline{M}(\partial T_j)\} < \infty .$$

Then, if $T_j \to T$ in $\mathcal{D}_\ell(U)$ for some $T \in \mathcal{D}_\ell(U)$ the limit T is in $\mathcal{R}_\ell(U)$, moreover $\partial T \in \mathcal{R}_{\ell-1}(U)$.

Remark:

1. We always have $T_j \to T$ in $\mathcal{D}_\ell(U)$ at least for a subsequence on account of Lemma 2.2.1.

2. A proof can be found in Simon's textbook, [74, Theorem 32.2].

Optimal Isoperimetric Inequality (Almgren 1986): Suppose that $S \in \mathcal{R}_{\ell-1}(\mathbb{R}^{m+k})$ is a closed $(\ell-1)$–current (i.e. $\partial S = 0$) of finite mass. Then there exists a mass minimizing current $Q \in \mathcal{R}_\ell(\mathbb{R}^{m+k})$ with boundary $\partial Q = S$ and

$$\underline{M}(Q) \leq \gamma_\ell \cdot \underline{M}(S)^{\ell/\ell-1} ,$$

$$\gamma_\ell := \ell^{-1-1/(\ell-1)} \cdot \alpha_\ell^{-1/(\ell-1)} .$$

Equality holds if and only if Q is an ℓ–dimensional ball with multiplicity $\Theta \equiv \text{const} \in \mathbb{N}$.

Comments:

1. The existence of a current $Q \in \mathcal{R}_\ell(\mathbb{R}^{m+k})$ minimizing mass with respect to the boundary S is rather easy to establish: it is sufficient to produce one element T in $\mathcal{R}_\ell(\mathbb{R}^{m+k})$ with finite mass and $\partial T = S$. Then the compactness theorem applied to a minimizing sequence gives a solution. A current T having the desired properties is the so-called *join* $0 \ast S$ which can be seen as an oriented cone over S with vertex 0.

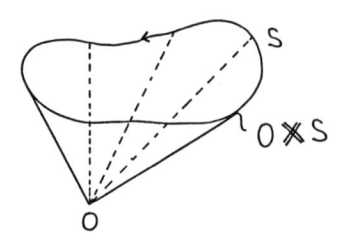

2. The Deformation Theorem implies
$$\underline{M}(Q) \leq c(m, k, \ell) \, \underline{M}(S)^{\frac{\ell}{\ell-1}}$$
with a rather large constant c.

3. Recently, Almgren [3] proved that the optimal isoperimetric constant γ_ℓ (known only for hypersurfaces S!) is independent of the codimension. Since in more than one codimension the boundary S can be filled in various ways it is natural that "the" isoperimetric inequality can only hold for minimizing currents Q.

4. As we shall see below our existence theorems for the M.C.P. are directly related to the optimal isoperimetric constant γ_ℓ. Moreover, we make use of the fact that "=" only occurs in the case of balls with constant multiplicity. □

Let us now return ot the M.C.P. The approach outlined here and in the next sections is due to Duzaar and the author. Our results are taken from the papers [10], [13], [15] – [17], [34], and [35].

For the rest of this section suppose that we are given an oriented mean curvature form $\Omega : \mathbb{R}^{m+k} \times \Lambda_m \mathbb{R}^{m+k} \to \mathbb{R}^{m+k}$ with associated scalar mean curvature form
$$\omega : \mathbb{R}^{m+k} \times \Lambda_{m+1} \mathbb{R}^{m+k} \to \mathbb{R}.$$
We assume that ω is a *bounded Borel function* being closed in the sense that

2.2 Some Facts from Geometric Measure Theory

$$\begin{cases} \partial Q(\omega) = 0 \\ \forall\, Q \in \mathcal{R}_{m+2}, \quad \underline{M}(Q) + \underline{M}(\partial Q) < \infty \end{cases} \quad (2.2.4)$$

Note that $Q \in \mathcal{R}_{m+2}$ together with $\underline{M}(\partial Q) < \infty$ implies $\partial Q \in \mathcal{R}_{m+1}$, i.e. $\partial Q = \underline{\tau}(M, \xi, \Theta)$ and $\partial Q(\omega) = 0$ just means

$$\int_M \langle \omega, \xi \rangle \Theta \, d\mathcal{H}^{m+1} = 0\,;$$

the integral is well defined on account of the properties of ω. In the case that ω is of class C^1 (2.2.4) is equivalent to the closedness of ω in the usual sense, i.e. $d\omega = 0$.

Let Γ denote a current in $\mathcal{R}_{m-1}$ s.t.

$$\partial \Gamma = 0, \quad \underline{M}(\Gamma) < \infty\,.$$

Definition: A current $T \in \mathcal{R}_m$ is a solution of the M.C.P. with curvature form Ω and boundary Γ iff $\underline{M}(T) < \infty$, $\partial T = \Gamma$ and $(T = \underline{\tau}(M, \xi, \Theta))$

$$\int_M \mathrm{div}_M X \cdot d\mu_T + \int_\Omega X \cdot \Omega(\xi) \, d\mu_T = 0 \quad (2.2.5)$$

for all $X \in C_0^1(\mathbb{R}^{m+k}, \mathbb{R}^{m+k})$, $\mathrm{spt}\, X \cap \mathrm{spt}\, \Gamma = \emptyset$. $(\mu_T := \mathcal{H}^m \llcorner \Theta)$

Remarks:

1. In case $T = [\![\Sigma]\!]$ for a smooth m–manifold Σ (2.2.5) clearly reduces to our previous formulation of the M.C.P.

2. As already remarked we will try to solve (2.2.5) by minimizing a suitable functional in the space $\{T \in \mathcal{R}_m : \underline{M}(T) < \infty,\ \partial T = \Gamma\}$. On the other hand (2.2.5) makes sense for objects $V = \underline{v}(M, \Theta)$ with M locally m–rectifiable and $\Theta : M \to \mathbb{N}$ locally $\mathcal{H}^m$–integrable provided we replace Ω by a mapping

$$\begin{cases} \Omega^* : \mathbb{R}^{m+k} \times G_m(\mathbb{R}^{m+k}) \to \mathbb{R}^{m+k}, \\ \Omega^*(x, E) \in E^\perp\,. \end{cases}$$

A pair $V = \underline{v}(M, \Theta)$ is called an m–*varifold* (compare [1]) and V has prescribed mean curvature Ω^* if (2.2.5) holds with $\Omega_p(\xi(P))$ replaced by $\Omega_p^*(T_pM)$. This would be a meaningfull concept for the unoriented M.C.P. but unfortunately the boundary of a varifold cannot be defined in an obvious way so that we have no variational setting.

3. Another extension of (2.2.5) would be to consider functions $\Omega : \mathbb{R}^{m+k} \times \Lambda_m \mathbb{R}^{m+k} \to \mathbb{R}^{m+k}$ with nonlinear dependence of the second argument. But in this case we failed to obtain a variational formulation.

$\square$

As a first step for solving (2.2.5) we introduce a functional whose first variation is just the mean curvature operator:

Let $T_0 \in \mathcal{R}_m$ denote a current of finite mass s.t. $\partial T_0 = \Gamma$. (The existence of T_0 follows for example from the Isomperimetric Theorem.) We think of T_0 as an arbitrary but fixed reference configuration in the class

$$\mathcal{C} := \{T \in \mathcal{R}_m : \underline{M}(T) < \infty,\ \partial T = \Gamma\}$$

and define

$$\boxed{\mathcal{F} : \mathcal{C} \to \mathbb{R}, \quad \mathcal{F}(T) := \underline{M}(T) + V_\omega(T, T_0)}$$

where

$$\boxed{V_\omega(T, T_0) := Q(\omega)}$$

for an arbitrary $(m+1)$-current $Q \in \mathcal{R}_{m+1}$, $\underline{M}(Q) < \infty$, $\partial Q = T - T_0$. From (2.2.4) we deduce that $V_\omega(T, T_0)$ is independent of the special Q, for example we may choose "the" mass minimzer for the boundary $T - T_0$. $V_\omega(T, T_0)$ is called the ω–volume enclosed by the currents T and T_0. In the codimension 1 case one should think of $V_\omega(T, T_0)$ as the volume of the "body" enclosed by the "surfaces" T and T_0 where the volume is measured with a weight function ω. If T_0' is in $\mathcal{C}$ then

$$V_\omega(T, T_0') = V_\omega(T, T_0) + V_\omega(T_0, T_0')$$

so that the functionals $\mathcal{F}$ and $\mathcal{F}'(T) := \underline{M}(T) + V_\omega(T, T_0')$ differ by an additive constant. This shows that the choice of the reference configuration T_0 is of no importance.

The motivation for introducing the $\mathcal{F}$–functional is the classical 2–dimensional parametric setting where one also introduces "the algebraic volume" enclosed by $u(D)$ and a fixed reference surface with the same boundary curve. After a short calculation this volume if of the type $V_H(u_\#([\![D]\!]), u_{0\#}([\![D]\!]))$ and once having obtained this expression our definition of $V_\omega(T, T_0)$ seems to be natural. Moreover, for conformal $u : D \to \mathbb{R}^3$ the mass of the current $u_\#([\![D]\!])$ is $\frac{1}{2}\int_D |\nabla u|^2\,dx$. Suppose that ω is C^1 so that $d\omega = 0$. Then there exists an m–form α with the property $d\alpha = \omega$ and we obtain at least formally

2.2 Some Facts from Geometric Measure Theory

$$\begin{aligned} V_\omega(T, T_0) &= Q(\omega) = Q(d\alpha) \\ &= \partial Q(\alpha) = T(\alpha) - T_0(\alpha) \end{aligned}$$

which suggests that it would be more explicit to discuss the functional

$$\tilde{\mathcal{F}}(T) = \underline{M}(T) + T(\alpha).$$

But if for example Ω is constant, i.e. independent of the base point, then $|\alpha_x|$ is of linear growth in x so that $T(\alpha)$ not necessarily makes sense for $T \in \mathcal{C}$. In order to have $T(\alpha)$ well defined spt T must be compact which is a severe restriction. Roughly speaking, the α–approach makes sense (and is very easy) whenever it is possible to construct α with

$$|\alpha| \leq 1 - \varepsilon$$

for some $\varepsilon > 0$ since in this case

$$\mathcal{F}(T) \geq \varepsilon \cdot \underline{M}(T)$$

which immediately implies that

$$\tilde{\mathcal{F}} \to \text{Min} \ \ \text{in} \ \mathcal{C}$$

has a solution. For example if spt $\Gamma \subset \overline{B}_R^{m+k}(x_0)$ and if

$$\sup_{|x| \leq R} |\omega_x| < m/R$$

then the ε–condition holds (on $B_R^{m+k}(x_0)$) and leads to an $\tilde{\mathcal{F}}$–minimizer with support in this ball.

We have discussed the above curvature condition in section 2.1 and recognized that bounds of the type

$$\left\{ \begin{array}{l} |\Omega| = |\omega| < c_m \cdot \underline{M}(\Sigma_0)^{-1/m}, \\ \Sigma_0 \text{ a mass minimizer for the boundary } \Gamma \end{array} \right\}$$

lead to larger admissible curvatures. As we shall see in the next section the discussion of the weighted volume $V_\omega(T, T_0)$ using Almgren's Optimal Isoperimetric Theorem will give existence under the above weaker condition.

Next we state the *first variation formula*: Consider a current $T \in \mathcal{C}$ and a 1–parameter family

$$\phi(t, \cdot) : \mathbb{R}^{m+k} \to \mathbb{R}^{m+k}, \quad -\varepsilon \leq t \leq \varepsilon,$$

of diffeomorphisms such that
$$\phi(0,\cdot) = \mathcal{I}d$$
and
$$\phi(t,x) = x, \quad |t| \le \varepsilon, \; x \in \mathbb{R}^{m+k} - K$$
for a compact subset $K \subset \mathbb{R}^{m+k}$ such that
$$K \cap \operatorname{spt} \Gamma = \emptyset.$$
A typical example is
$$\phi(x,t) := x + t \cdot X(x), \quad X \in C_0^1(\mathbb{R}^{m+k}, \mathbb{R}^{m+k}), \; \operatorname{spt} X \cap \operatorname{spt} \Gamma = \emptyset$$
for which $K = \operatorname{spt} X$.

The currents $\phi(t,\cdot)_\# T$ are in the class $\mathcal{C}$ and we want to calculate
$$\frac{d}{dt/0} \mathcal{F}(\phi(t,\cdot)_\# T) = \frac{d}{dt/0} \underline{M}(\phi(t,\cdot)_\# T) + \frac{d}{dt/0} V_\omega(\phi(t,\cdot)_\# T, T_0).$$
Following for example [74] we observe the formula
$$\underline{M}(\phi(t,\cdot)_\# T) = \int_M J_M \phi(t,\cdot) \cdot \Theta \, d\mathcal{H}^m$$
($T := \underline{\tau}(M, \xi, \Theta)$, $J_M \phi(x,t) =$ relative Jacobian of $\phi(t,x) = (\det[d_M \phi(t,x)^* \circ d_M \phi(t,x)])^{1/2}$) and after some calculations we arrive at
$$\left| \begin{array}{l} \dfrac{d}{dt/0} \underline{M}(\phi(t,\cdot)_\# T) = \displaystyle\int_M \operatorname{div}_M X \cdot d\mu_t, \\[2mm] X = \dfrac{\partial}{\partial t/0} \phi(t,\cdot), \quad \mu_T = \mathcal{H}^m \, \llcorner \, \Theta \end{array} \right.$$
Next we observe
$$\begin{aligned} \frac{d}{dt/0} V_\omega(\phi(t,\cdot)_\# T, T_0) &= \frac{d}{dt/0} \{V_\omega(\phi(t,\cdot)_\# T, T) + V_\omega(T, T_0)\} \\ &= \frac{d}{dt/0} V_\omega(\phi(t,\cdot)_\# T, T) \\ &= \frac{d}{dt/0} Q_t(\omega) \end{aligned}$$
where Q_t is an arbitrary $(m+1)$–current bounding $\phi(t,\cdot)_\# T - T$.

We choose
$$Q_t := \phi_\#([\![0,t]\!] \times T)$$
where $[\![0,t]\!] \times T \in \mathcal{R}_{m+1}(\mathbb{R} \times \mathbb{R}^{m+k})$ is the Cartesian product of the current $[\![0,t]\!] \in \mathcal{R}_1(\mathbb{R})$ and the given current $T \in \mathcal{R}_m(\mathbb{R}^{m+k})$.

2.3 A First Approach to the Mean Curvature Problem

Using the formula (see [74, formula 26.22])

$$\phi_\#(\llbracket 0,t \rrbracket \times T)(\omega)$$

$$= \int_0^t \int_M \langle \omega(\phi(s,x)) , \frac{\partial}{\partial s}\phi(s,x) \wedge \Lambda D_x\phi(s,x)(\vec{T}(x))\rangle \, d\mu_T(x) \, ds$$

we immediately obtain

$$\frac{d}{dt/0} Q_t(\omega) = \int_M \langle \omega, X \wedge \vec{T} \rangle \, d\mu_T = \int_M \Omega(\vec{T}) \cdot X \, d\mu_T \, .$$

Lemma 2.2.2 *For any current $T \in \mathcal{C}$ and a family of diffeormorphisms $\phi(t,\cdot)$ as above we have*

$$\left| \frac{d}{dt/0} \mathcal{F}(\phi(t,\cdot)_\# T) \right.$$
$$= \int_M (div_M X + X \cdot \Omega(\vec{T})) \, d\mu_T \, ,$$

$$X = \frac{\partial}{\partial t/0} \phi(t,\cdot), \quad \mu_T = \mathcal{H}^m \, \llcorner \, \Theta, \quad T = \underline{\tau}(M,\vec{T},\Theta) \, .$$

Hence $0 = \frac{d}{dt/0} \mathcal{F}(\phi(t,\cdot)_\# T)$ for all $\phi(t,\cdot)$ of the form $\phi(t,x) = x + t \cdot X(x)$ with X arbitrary implies that T is a current solution of the M.C.P. In order to obtain such T we now discuss the problem of minimizing $\mathcal{F}$.

2.3 A First Approach to the Mean Curvature Problem

We use the notations introduced at the end of section 2.2 and try to find solutions $T \in \mathcal{C}$ of

$$\mathcal{F}(T) := \underline{M}(T) + V_\omega(T, T_0) \to \text{Min in } \mathcal{C}$$

which by Lemma 2.2.2 are generalized manifolds of mean curvature Ω. But without further assumptions the functional $\mathcal{F}$ is in general not bounded from below. So in a first step we replace $\mathcal{C}$ by the subclass

$$\mathcal{C}_R := \{T : \underline{M}(T) \leq R\} \cap \mathcal{C}$$

where $R \geq \inf\{\underline{M}(S) : S \in \mathcal{C}\}$ is an arbitrary real number. Clearly $\mathcal{C}_R \neq \emptyset$ and

$$\mathcal{F}(T) \geq -|V_\omega(T,T_0)| = -|Q(\omega)| \geq -|\omega| \cdot \underline{M}(Q)$$

where $Q \in \mathcal{R}_{m+1}$ is mass minimizing for the boundary $T - T_0$. The isoperimetric theorem implies

$$\underline{\underline{M}}(Q) \leq \gamma_{m+1} \cdot \underline{\underline{M}}(T - T_0)^{\frac{m+1}{m}} \leq \gamma_{m+1}(R + \underline{\underline{M}}(T_0))^{1+1/m}$$

so that

$$\mathcal{F}(T) \geq -|\omega| \cdot \gamma_{m+1} (R + \underline{\underline{M}}(T_0))^{1+1/m},$$

hence

$$\inf_{\mathcal{C}_R} \mathcal{F} > -\infty$$

and we may consider an $\mathcal{F}$–minimizing sequence $\{T_i\} \in \mathcal{C}_R$. Since

$$\underline{\underline{M}}(T_i) \leq R(<\infty)$$

we have (after passing to a subsequence if necessary) $T_i \to T$ for some $T \in \mathcal{C}_R$. We expect $\mathcal{F}(T) = \inf_{\mathcal{C}_R} \mathcal{F}$. Consider mass minimizing currents $Q_i \in \mathcal{R}_{m+1}$ for the boundaries $T_i - T_0$. The isoperimetric theorem implies

$$\sup_{i \in \mathbb{N}} \underline{\underline{M}}(Q_i) < \infty$$

and therefore (after selecting a subsequence) $Q_i \to: Q \in \mathcal{R}_{m+1}$, $\underline{\underline{M}}(Q) < \infty$ and $\partial Q = T - T_0$. But

$$V_\omega(T_i, T_0) = Q_i(\omega) \underset{i \to \infty}{\to} Q(\omega) = V_\omega(T, T_0)$$

does in general *not* follow from current convergence $Q_i \to Q$ (or equivalently: the volume functional is *not* continuous in the sense that $T_i \to T$ implies $V_\omega(T_i, T_0) \to V_\omega(T, T_0)$).

Sufficient conditions for $Q_i(\omega) \to Q(\omega)$ are: Ω is *continuous* and satisfies

	(i)	spt ω is compact
or	(ii)	$\sup_{|z| \geq \rho} |\omega_z| \to 0$ as $\rho \to \infty$
or	(iii)	there is a compact set $K \subset \mathbb{R}^{m+k}$ such that spt $T_0 \cup$ spt $T_i \subset K$

For (i) one has to observe that in case of finite mass currents the current convergence $Q_i \to Q$ also implies $Q_i(\eta) \to Q(\eta)$ for continuous forms with compact support. In case (ii) we observe $Q_i(\varphi \cdot \omega) \underset{i \to \infty}{\to} Q(\varphi \cdot \omega)$ for any $\varphi \in C_0^0(\mathbb{R}^{m+k})$. Applying this to a function $0 \leq \varphi \leq 1$ with $\eta \equiv 1$ on B_ρ, $\rho \gg 1$, the result follows since the contribution

2.3 A First Approach to the Mean Curvature Problem

$$Q_i((1-\varphi)\omega)$$

is controlled by $\sup\limits_{|z|\geq\rho}|\omega_z|\cdot\underline{\underline{M}}(Q_i)$.

Let (iii) hold and consider the convex hull K' of K. Then $\operatorname{spt} Q_i$, $\operatorname{spt} Q \subset K'$ and we are in a situation similar to (i).

Therefore we now consider the following restricted variational problem: Suppose we are given

(i)   a continuous bounded mean curvature form Ω whose associated scalar mean curvature form is weakly closed

(ii) a compact set $K \subset \mathbb{R}^{m+k}$ and

(iii) a current $T_0 \in \mathcal{R}_m$ with $\operatorname{spt} T_0 \subset K$ and $\underline{\underline{M}}(T_0) < \infty$.

Let

$$\mathcal{C}_K := \{T \in \mathcal{R}_m : \partial T = \partial T_0 \,,\, \underline{\underline{M}}(T) < \infty \,,\, \operatorname{spt} T \subset K\}$$

and define

$$m_K := \inf\{\underline{\underline{M}}(S) : S \in \mathcal{C}_K\}\,.$$

For $R \geq m_K$ let us finally set

$$\mathcal{C}_K^R := \{T \in \mathcal{C}_K : \underline{\underline{M}}(T) \leq R\}\,.$$

Theorem 2.3.1 *Under the above hypothesis*

$$\mathcal{F}(T) = \underline{\underline{M}}(T) + V_\omega(T, T_0) \to \min \quad \text{in } \mathcal{C}_K^R$$

has at least one solution.

Remarks:

1. The prescribed boundary $\Gamma \in \mathcal{R}_{m-1}$ ($\underline{\underline{M}}(\Gamma) < \infty$, $\partial \Gamma = 0$) has now been replaced by ∂T_0 where we assume that there exists at least one surface T_0 with boundary Γ and staying inside K. Of course we can formulate geometric conditions on K implying the existence of such a T_0.

2. Our existence theorem does not necessarily provide us with a current solution of the M.C.P. for the reason that the class $\mathcal{C}_K^R$ is defined by nonlinear side conditions which are not respected by arbitrary variations.

So we have to exhibit geometric conditions on R and K guaranteeing that solutions T produced in Theorem 2.3.1 satisfy

$$\underline{\underline{M}}(T) < R, \quad \text{dist}(\text{spt}\,T, \partial K) > 0\,.$$

We start with a *maximum-principle for the mass:* Consider an $\mathcal{F}$–minimizer T in $\mathcal{C}_K^R$ and a current S such that $m_K = \underline{\underline{M}}(S)$. Then

$$\mathcal{F}(T) \leq \mathcal{F}(S)\,,$$

hence

$$\begin{aligned}\underline{\underline{M}}(T) &\leq \underline{\underline{M}}(S) + V_\omega(S,T) \\ &\leq \underline{\underline{M}}(S) + \gamma_{m+1}\cdot|\omega|\cdot[\underline{\underline{M}}(T)+\underline{\underline{M}}(S)]^{1+1/m} \\ &\leq \underline{\underline{M}}(S) + \gamma_{m+1}\cdot|\omega|\cdot\underline{\underline{M}}(S)^{1+1/m}\cdot[1+R/\underline{\underline{M}}(S)]^{1+1/m}\,.\end{aligned}$$

Let $c := R/\underline{\underline{M}}(S)$ (≥ 1). The right-hand-side of the above inequality is strictly smaller than R if

$$1 + \gamma_{m+1}\cdot|\omega|\cdot\underline{\underline{M}}(S)^{1/m}\cdot[1+c]^{1+1/m} < c\,,$$

and this inequality clearly is equivalent to the curvature bound

$$|\omega| < (c-1)\cdot\gamma_{m+1}^{-1}\cdot\underline{\underline{M}}(S)^{-1/m}[1+c]^{-1-1/m} =: f(c) \qquad (2.3.1)$$

and in order to get large admissible curvatures ω we have to maximize $f(c)$ on $[1,\infty)$. A short computation yields $c = 2m+1$ and (2.3.1) implies

Theorem 2.3.2 *Suppose that the hypothesis of Theorem 2.3.1 are satisfied with*

$$R := (2m+1)\cdot m_K \qquad (2.3.2)$$

and suppose the curvature bound

$$|\omega| < \gamma_{m+1}^{-1}\, 2^{-1/m}\cdot m\cdot(m+1)^{-1-1/m}\, m_K^{-1/m}\,. \qquad (2.3.3)$$

Then, any $\mathcal{F}$–minimizer in $\mathcal{C}_K^R$ satisfies $\underline{\underline{M}}(T) < R$, i.e. T is an interior minimizer w.r.t. $\underline{\underline{M}}$. □

We remark (compare section 2.4 for details) that conditions (2.3.2), (2.3.3) are sufficient for proving the existence of minimizers in the larger class $C^R = \{T \in \mathcal{R}_m : \underline{\underline{M}}(T) \leq R,\ \partial T = \partial T_0\}$ where the condition $\text{spt}(\cdot) \subset K$ is dropped. (The number m_K has to be replaced by $\inf\{\underline{\underline{M}}(S) : \partial S = \partial T_0\}$.) Since the conclusion of Theorem 2.3.2 continues to hold the minimizer then solves the M.C.P.

2.3 A First Approach to the Mean Curvature Problem

We now discuss conditions on K implying that $\mathcal{F}$-minimzers do not contact ∂K. Let us assume that K is the (compact) *closure of a C^2-domain* in $\mathbb{R}^{m+k}$ with interior unit normal vector field $\mathcal{N}$. Consider $X \in C_0^1(\mathbb{R}^{m+k}, \mathbb{R}^{m+k})$, $\operatorname{spt} X \cap \operatorname{spt} \partial T_0 = \emptyset$, and suppose that $\operatorname{spt} X$ is concentrated in a small ball B centered at x_0. If *(2.3.2) and (2.3.3)* hold and if also B is contained in the interior of K then for $|t| \ll 1$

$$\phi(t, \cdot)_{\#} T \in \mathcal{C}_K^R, \quad \phi(t, x) = x + t \cdot X(x),$$

hence

$$\frac{d}{dt/0} \mathcal{F}(\phi(t, \cdot)_{\#} T) = 0.$$

The situation becomes more complicated in the case $x_0 \in \partial K$. Then, similar to the proof of Theorem 1.2.1 in chapter 1 we consider normal and tangential variations seperately:

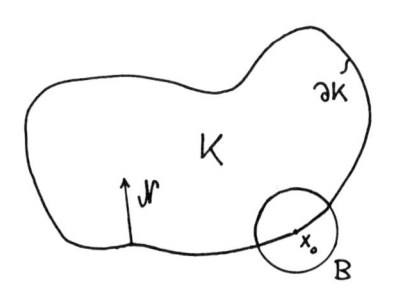

$d := \operatorname{dist}(\cdot, \partial K)$ is smooth in a onesided neighborhood of ∂K and we let $\mathcal{N}^* = \operatorname{grad} d$ there, so that $\mathcal{N}^* = \mathcal{N}$ on ∂K.

Define

$$\phi(s, x) := x + s \cdot \eta(x) \cdot h_\varepsilon(d(x)) \mathcal{N}^*(x),$$

$$h_\varepsilon : [0, \infty) \to [0, 1], \quad h_\varepsilon = 1 \text{ on } [0, \varepsilon],$$

$$h_\varepsilon = 0 \text{ on } [2 \cdot \varepsilon, \infty], \quad h'_\varepsilon \leq 0,$$

$$s > 0, \quad \eta \in C_0^1(B), \quad \eta \geq 0$$

and observe

$$T_s := \phi(s, \cdot)_{\#} T \in \mathcal{C}_K^R$$

for s small (clearly $\underline{M}(T) < R$ gives $\underline{M}(T^s) \leq R$). From $\frac{1}{s}(\mathcal{F}(T_s) - \mathcal{F}(T)) \geq 0$ we infer

$$\int \eta \, d\lambda = \int_M \left(\operatorname{div}_M \left(\eta \cdot h_\varepsilon(d) \mathcal{N}^* \right) + \eta h_\varepsilon(d) \mathcal{N}^* \cdot \vec{\Omega}(\vec{T}) \right) d\mu_T$$

for a Radon measure $\lambda \geq 0$ independent of ε for which we obtain the estimate

$$\begin{cases} \int \eta \, d\lambda \leq \int_{M \cap \partial K} \{\mathcal{N} \cdot \Omega(\vec{T}) + \mathrm{div}_M \, \mathcal{N}\} \cdot \eta \cdot d\mu_T \,, \\ \text{moreover } \{\ldots\} \geq 0 \quad \mu_T - \text{a.e. on } M \cap \partial K \,. \end{cases}$$

Hence $\lambda = \mu_T \llcorner \vartheta \cdot \{\ldots\}$ for a density $\vartheta : M \cap \partial K \to [0,1]$. Next consider a field $Y \in C_0^1(B, \mathbb{R}^{m+k})$ being tangential to ∂K, i.e. $Y \cdot \mathcal{N}^* = 0$, and its associated global flow $\phi(t,x)$. We define

$$\psi(s,x) := \phi(s \cdot \eta \cdot h_\varepsilon(d), x), \quad |s| \ll 1, \, \eta \in C_0^1(B),$$

and observe $\psi(s,\cdot)_\# T \in \mathcal{C}_K^R$ which gives

$$0 = \int_M \left(\mathrm{div}_M \left(\eta \cdot h_\varepsilon(d) \cdot Y \right) + \eta \cdot h_\varepsilon(d) Y \cdot \Omega(\vec{T}) \right) d\mu_T \,.$$

Putting together these equations we arrive at

Theorem 2.3.3 *Suppose that (2.3.2), (2.3.3) hold and that T is an $\mathcal{F}$-minimizer in $\mathcal{C}_K^R$ with K as above. Then for any $X \in C_0^1(\mathbb{R}^{m+k}, \mathbb{R}^{m+k})$, $\mathrm{spt}\, X \cap \mathrm{spt}\, \partial T_0 = \emptyset$,*

$$\begin{cases} \int_M \left(\mathrm{div}_M \, X + X \cdot \Omega(\vec{T}) \right) d\mu_T \\ = \int_{M \cap \partial K} \vartheta X \cdot N \cdot \{\mathrm{div}_M \, \mathcal{N} + \mathcal{N} \cdot \Omega(\vec{T})\} \, d\mu_T \end{cases} \quad (2.3.4)$$

with $\{\ldots\} \geq 0$ μ_T − a.e. on $M \cap \partial K$ and a density $\vartheta : M \cap \partial K \to [0,1]$.

Now, what are the conditions on K implying that the right-hand side of equation (2.3.4) vanishes?

Let $\mathcal{K}_1(P), \ldots, \mathcal{K}_{m+k-1}(P)$, $P \in \partial K$, denote the *principal curvatures* of ∂K at P calculated w.r.t. the interior normal $\mathcal{N}$. We abbreviate

$$\mathcal{K}(P) := \min\{\mathcal{K}_i(P) : i = 1, \ldots, m+k-1\} \,.$$

We know

$$\mathrm{div}_M \, \mathcal{N} + \mathcal{N} \cdot \Omega(\vec{T}) \geq 0$$

a.e. on $M \cap \partial K$. For $P \in \partial K \cap M$ let $\tau_1, \ldots, \tau_m$ denote an ONB in $T_p M \subset T_p \partial K$ with $\vec{T}(P) = \tau_1 \wedge \ldots \wedge \tau_m$. Then

$$-\mathrm{div}_M \, \mathcal{N}(P) = \sum_{i=1}^m -\tau_i \cdot \partial_{\tau_i} \mathcal{N}(P) \geq m \cdot \mathcal{K}(P)$$

2.3 A First Approach to the Mean Curvature Problem

since $-\tau \cdot \partial_\tau \mathcal{N}(P) \geq \mathcal{K}(P)$, $\forall \tau \in T_p \partial K$, $|\tau| = 1$. This implies

$$\text{div}_M \mathcal{N}(P) + \mathcal{N}(P) \cdot \Omega_p(\vec{T}(P)) \leq -m \cdot \mathcal{K}(P) + |\Omega|$$

and the above calculations hold for a.a. $P \in M \cap \partial K$. So, if we require

$$|\Omega_p| \leq m \cdot \mathcal{K}(P), \quad \forall P \in \partial K, \tag{2.3.5}$$

then $\text{div}_M \mathcal{N} + \mathcal{N} \cdot \Omega(\vec{T}) \leq 0$ on the contact set $M \cap \partial K$ and we arrive at

$$\text{div}_M \mathcal{N} + \mathcal{N} \cdot \Omega(\vec{T}) = 0 \quad \text{a.e.}$$

Theorem 2.3.4 (existence of solutions of the M.C.P.) *Suppose that we are given a bounded continuous mean curvature form ω which is weakly closed. Let K denote the closure of a C^2-domain with inward normal $\mathcal{N}$ and principal curvatures $\mathcal{K}_i$ (w.r.t. $\mathcal{N}$). Consider a current T_0 of finite mass and $\text{spt}\, T_0 \subset K$. Define the minimal area*

$$m_K := \inf\{\underline{M}(S) : S \in \mathcal{R}_m, \partial S = \partial T_0, \text{spt}\, S \subset K\}$$

and suppose

$$|\omega_p| = |\Omega_p| \leq m \cdot \min_\ell \mathcal{K}_\ell(P), \quad P \in \partial K,$$

$$|\omega| = |\Omega| < \gamma_{m+1}^{-1} \cdot 2^{-1/m}\, m \cdot (m+1)^{-1-1/m}\, m_K^{-1/m}.$$

Then the M.C.P. has a solution $T = \underline{\tau}(M, \xi, \Theta)$, i.e.

$$\int_M \left(\text{div}_M X + X \cdot \Omega(\xi)\right) \Theta \, d\mathcal{H}^m = 0$$

for all $X \in C_0^1(\mathbb{R}^{m+k}, \mathbb{R}^{m+k})$, $\text{spt}\, X \cap \text{spt}\, \partial T_0 = \emptyset$, and $\partial T = \partial T_0$. □

Remark: The curvature condition (2.3.5) relating Ω to the principal curvatures of ∂K does not exclude the possibility that $\text{spt}\, T$ contacts the boundary ∂K. But if contact occurs then the curvature of ∂K is related to Ω through the formula

$$\sum_{i=1}^m (-\tau_i) \cdot \partial_{\tau_i} \mathcal{N} = \mathcal{N} \cdot \Omega(\tau_1 \wedge \ldots \wedge \tau_m).$$

□

Following an argument of [51] obtained for parametric mean curvature surfaces one can prove the following result:

Theorem 2.3.5 *Suppose that the hypothesis of Theorem 2.3.4 hold but with the stronger curvature bound*

$$|\omega_p| = |\Omega_p| < m \cdot \min_\ell \mathcal{K}_\ell(P), \quad P \in \partial K.$$

Assume further that $\operatorname{dist}(\operatorname{spt}\partial T_0, \partial K) > 0$. *Then, if* $T \in \mathcal{C}_K^R$ *denotes an* $\mathcal{F}$-*minimizing current in* $\mathcal{C}_K^R$ *we have* $\operatorname{dist}(\operatorname{spt} T, K) > 0$.

Remark: Since we already know $\underline{M}(T) < R$ we deduce from Theorem 2.3.5 that T solves the M.C.P. without referring to the Euler equation obtained in Theorem 2.3.3.

The idea for *the proof of Theorem 2.3.5* is the following: Again let $d := \operatorname{dist}(\cdot, \partial K)$ and define

$$K_\rho := \{x \in K : d(x) < \rho\}.$$

For ρ small

$$\Pi : K_\rho \to \partial K, |\Pi(x) - x| = d(x),$$
(the nearest point projection)

is well defined and smooth. Finally let

$$P_\rho(x) := \begin{cases} x, & d(x) \geq \rho \\ \Pi(x) + \rho \cdot \mathcal{N}(\Pi(x)) \end{cases}, x \in K.$$

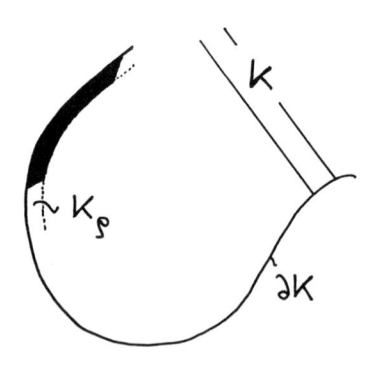

P_ρ is the identity map for x having distance $\geq \rho$ to ∂K, all other points are mapped on the set $[d = \rho]$. From $\operatorname{dist}(\operatorname{spt}\partial T_0, \partial K) > 0$ we deduce $(P_\rho)_\# T \in \mathcal{C}_K^R$ and since T is minimal $\mathcal{F}(T) \leq \mathcal{F}((P_\rho)_\# T)$. On the other hand – using the strong curvature assumption –

$$\mathcal{H}^m(\operatorname{spt} T \cap K_\rho) > 0$$

for all $\rho > 0$ implies $\mathcal{F}((P_\rho)_\# T) - \mathcal{F}(T) < 0$, i.e. squeezing T via P_ρ decreases energy contradicting the minimality of T. □

We close this section with some *comments concerning the regularity of our solutions to the M.C.P.*

2.3 A First Approach to the Mean Curvature Problem

It is well known that even for mass minimizing currents (corresponding to the case $\Omega \equiv 0$) singular solutions may occur so that the best possible results one can hope for are the partial regularity theorems obtained for mass minimizers.

So let T denote a current solution of the M.C.P. obtained for example in Theorem 2.3.4 or 2.3.5. We let

$$\operatorname{Reg}(T) := \{z \in \mathbb{R}^{m+k} - \operatorname{spt}(\partial T_0) : \text{there is a ball } B_r(z) \text{ such} \\ \text{that } \operatorname{spt} T \cap B_r(z) \text{ is an } m\text{-dimensional oriented} \\ \text{smooth manifold with mean curvature form } \Omega\}$$

denote the *interior regular set* of T. We consider the associated varifold $V = \underline{v}(M, \Theta)$ ($T = \underline{\tau}(M, \xi, \Theta)$, $\mu_V := \mathcal{H}^m \mathbin{\llcorner} \Theta$) and observe that the mean curvature equation implies

$$\int_U \operatorname{div}_M X \cdot d\mu_V = -\int_U X \cdot F \, d\mu_V$$

$$\forall X \in C_0^1(\mathbb{R}^{m+k}, \mathbb{R}^{m+k}), \ \operatorname{spt} X \subset U,$$

U open and disjoint to $\operatorname{spt} \partial T_0$.

Here $F(x) := \Omega_x(\xi(x))$.

Varifolds $V = \underline{v}(M, \Theta)$ for which such an equation holds on an open set $U \subset \mathbb{R}^{m+k}$ with a function $F \in L^p_{\text{loc}}(U, \mu_V)$ were studied by Allard [1] (compare also [74, chapter 5]) and from Allard's work we immediately obtain

Theorem 2.3.6 (Allard regularity) *If $T \in \mathcal{R}_m$ solves the M.C.P. then $\operatorname{Reg}(T)$ is open and dense in $\operatorname{spt} T - \operatorname{spt} \partial T_0$. More precisely: There exists a number $\delta > 0$ such that*

$$\alpha_m^{-1} r^{-m} \int_{M \cap B_r(z)} \Theta \cdot d\mathcal{H}^m \leq 1 + \delta$$

for some ball $B_r(z)$, $z \in \operatorname{spt} T - \operatorname{spt} \partial T_0$, implies $z \in \operatorname{Reg} T$. □

The proof of this result makes strong use of monotonicity formulas for the quantity

$$\alpha_m^{-1} r^{-m} \int_{M \cap B_r(z)} \Theta \, d\mathcal{H}^m$$

which measures how close the area of T in the $(m+k)$–dimensional ball $B_r(z)$ is to $\alpha_m \cdot r^m$.

It should also be noted that Theorem 2.3.6 is just the application of the Allard regularity theorem (which holds for any varifold of bounded first variation), we do not make use of the fact that the current solution is obtained by minimizing. Perhaps minimality of T gives stronger information on $\operatorname{Reg} T$. For example Almgren [2] proves on 1600 pages the following result:

> If $S \in \mathcal{R}_m$ is *mass minimizing* then
> $$\mathcal{H} - \dim(\operatorname{Sing} S) \leq m - 2$$
> for the interior singular set.

In case $m = 2 = k$
$$\{(z, w) \in \mathbb{C}^2 = \mathbb{R}^4 : z^3 = w^2\}$$
is locally minimizing with singularity at 0 so that the upper bound $m - 2$ for the Hausdorff-dimension is optimal. Maybe that Almgren's theorem also holds for $\mathcal{F}$-minimizers.

In the *codimension 1 case* Theorem 2.3.6 can be improved considerably leading to the best possible regularity results. For $k = 1$ any current $T \in \mathcal{R}_m$ has a decomposition
$$T = \sum_{j=-\infty}^{\infty} \partial[\![E_j]\!]$$
with sets E_j of finite perimeter [74, Theorem 27.6] (i.e. $\mathbf{1}_{E_j} \in BV(\mathbb{R}^{m+1})$) which reduces the situation to the case $T = \partial[\![E]\!]$. Now we can argue as in [74, §37] with $\underline{M}$ replaced by $\mathcal{F}$ to get the following

Theorem 2.3.7 (interior regularity for $k = 1$) *Suppose $k = 1$ and that $T \in \mathcal{C}_K^R$ is $\mathcal{F}$-minimizing under the conditions of Theorem 2.3.4 or 2.3.5. Then*

> $\operatorname{Sing} T = \emptyset$ *if* $m \leq 6$,
> $\operatorname{Sing} T$ *discrete*   *for* $m = 7$,
> $\mathcal{H} - \dim(\operatorname{Sing} T) \leq m - 7$ *for* $m \geq 8$

hold for the interior singular set.

Remarks:

1. The above result is well known for mass minimizing currents and can in general not be improved because there exist 7–dimensional area minimizing surfaces in $\mathbb{R}^8$ with isolated singularities.

2. It is worth noting that Theorem 2.3.7 makes strong use of the *existence of area minimizing tangent cones* at every point $x \in \operatorname{spt} T - \operatorname{spt}(\partial T_0)$ whose existence is proved with the help of the monotonicity formula. It is then possible to apply Federer's dimension reduction argument: blowing up T at x (i.e. considering $T_\lambda := (\eta_\lambda)_\# T$, $\eta_\lambda(z) = \frac{1}{\lambda}(z-x)$, $\lambda \downarrow 0$) leads to a locally area minimizing m–dimensional cone in $\mathbb{R}^{m+1}$. In case $m \leq 6$ the cone must be a plane (see [75]) so that T has a tangent plane at x which gives $x \in \operatorname{Reg} T$.

Up to now we have been concerned with interior regularity. Very recently Duzaar & Steffen [19] proved the following:

> If $k = 1$ and if ∂T_0 is a smooth manifold then under the assumptions of Theorem 2.3.4 or 2.3.5 T is smooth in a tubular neighborhood of $\operatorname{spt} \partial T_0$ and $\operatorname{spt} \partial T_0$ is geometric boundary of $\operatorname{spt} T$.

This gives a complete description of the situation for $k = 1$ and $m \leq 6$: *In this case the M.C.P. admits classical solutions provided we impose natural curvature restrictions.*

2.4 General Existence Theorems, Applications to Isoperimetric Problems

If one starts with a boundary configuration Γ and a curvature form Ω it is somewhat artificial to construct first a set K containing Γ and then to look for the solution of the M.C.P. satisfying $\operatorname{spt}(\cdot) \subset K$: During this process one is faced with the difficulty that Ω and the principal curvatures of ∂K have to be related in the right way. The reason for introducing the set K and the additional support condition was that it is very easy to obtain $\mathcal{F}$–minimizing currents since then the volume term is continuous w.r.t. current convergence.

In this section we want to show how to remove the support condition by the way obtaining $\mathcal{F}$–minimizing currents under rather natural and (as we conjecture) optimal conditions.

Theorem 2.4.1 *Suppose that the given curvature form Ω is constant, i.e. just a linear function $\Lambda_m \mathbb{R}^{m+k} \to \mathbb{R}^{m+k}$ such that $\Omega(\tau_1 \wedge \ldots \wedge \tau_m) \in \operatorname{Span}[\tau_1, \ldots, \tau_m]^\perp$. Let $T_0 \in \mathcal{R}_m$ denote a current of finite mass with compact support and fix $R \geq \inf\{\underline{\underline{M}}(S) : S \in \mathcal{R}_m, \partial S = \partial T_0\} =: \lambda$. Then there is a solution T of*

$$\mathcal{F} = \underline{M}(\cdot) + V_\omega(\cdot, T_0) \to \min$$

in $\mathcal{C}_R := \{S \in \mathcal{R}_m : \underline{M}(S) \leq R, \partial S = \partial T_0\}$ *with compact support. If we let* $R := (2m+1) \cdot \lambda$ *and impose*

$$|\Omega| < \gamma_{m+1}^{-1} \cdot 2^{-1/m} \cdot m \cdot (m+1)^{-1-1/m} \lambda^{-1/m} \qquad (2.4.1)$$

then T is a solution of the M.C.P. for the boundary ∂T_0 and mean curvature form Ω. □

In case of constant curvature it is possible to minimize $\mathcal{F}$ in any class of currents with uniform mass bound and common boundary ∂T_0. There are *no* additional assumptions on the size of Ω. Condition (2.4.1) comes into play (through the maximum principle for the mass, Theorem 2.3.2) if we want our minimizer T to solve the mean curvature equation.

Sketch of the Proof (see [15]):

1. For $j \gg 1$ let

$$\mathcal{C}_j := \{T \in \mathcal{R}_m : \operatorname{spt} T \subset \overline{B}_j\} \cap \mathcal{C}_R$$

and by Theorem 2.3.1 we find $T_j \in \mathcal{C}_j$ s.t.

$$\mathcal{F}(T_j) = \inf_{\mathcal{C}_j} \mathcal{F}.$$

2. Define the *slice* $\langle Q, r \rangle$ of a current $Q \in \mathcal{R}_{m+1}$ at radius r through

$$\langle Q, r \rangle := \partial[Q \llcorner B_r] - (\partial Q) \llcorner B_r$$

(where $(S \llcorner A)(\phi) := \int_A \langle \phi, \vec{S} \rangle d\mu_s$ is the restriction). We have $\langle Q, r \rangle \in \mathcal{R}_m$ for almost all $r > 0$.

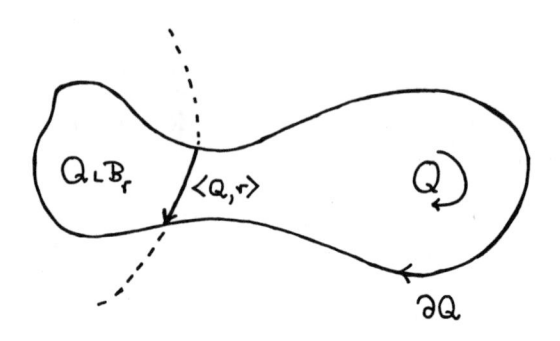

2.4 General Existence Theorems, Applications to Isoperimetric Problems

Key Lemma: *Let $Q_j \in \mathcal{R}_{m+1}$ denote a mass minimizer for the boundary $T_j - T_0$. Then there are numbers $0 < \rho < \sigma < \infty$ (independent of j) and a sequence $\{r_j\}$ of slicing radii such that*

$$\rho \leq r_j \leq \sigma, \quad \langle Q_j, r_j \rangle \in \mathcal{R}_m, \quad \underline{M}(\langle Q_j, r_j \rangle) \leq j^{-1}.$$

3. Let $T_j^* := T_j \llcorner B_{r_j} + \langle Q_j, r_j \rangle$. We have

$$\partial T_j^* = \partial T_0 \quad \text{and} \quad \operatorname{spt} T_j^* \subset \overline{B}_\sigma.$$

Clearly we may assume $\Omega \neq 0$. Then we find $w_1, \ldots, w_{m+1}$, $w_i \cdot w_j = \delta_{ij}$, such that ($\omega$ = scalar mean curvature form)

$$\omega(w_1 \wedge \ldots \wedge w_{m+1}) = |\omega|.$$

Let $E = \operatorname{Span}[w_1, \ldots, w_{m+1}]$, $[\![E]\!] = \underline{\tau}(E, w_1 \wedge \ldots \wedge w_m, 1)$ and define for $t > 0$, $\kappa \in \{-1, 1\}$

$$Y_{t,\kappa} := \underline{\tau}(E \cap B_t(0), \kappa \cdot \vec{E}, 1)$$

which is an oriented $(m+1)$–ball of radius t in $\mathbb{R}^{m+k}$. We have

$$V_\omega(\partial Y_{t,\kappa}, 0) = Y_{t,\kappa}(\omega) = \alpha_{m+1} \cdot \kappa \cdot |\omega| \, t^{m+1},$$

$$|Y_{t,\kappa}(\omega)| = \gamma_{m+1} |\omega| \cdot \underline{M}(\partial Y_{t,\kappa})^{1+1/m}.$$

We now choose $t_j > 0$, $\kappa_j \in \{1, -1\}$ to satisfy

$$V_\omega(\partial Y_j, 0) = -V_\omega(T_j^*, T_j), \quad Y_j := Y_{t_j, \kappa_j}$$

and an easy calculation shows

$$\sup_j t_j < \infty,$$

hence the currents

$$S_j := T_j^* + \partial Y_j$$

have *uniformly bounded supports*.

4. We claim

$$\underline{M}(S_j) \leq \underline{M}(T_j) + \frac{2}{j}, \quad \mathcal{F}(S_j) \leq \mathcal{F}(T_j) + \frac{2}{j}.$$

(Note $\partial S_j = \partial T_j^* = \partial T_0$ and $\operatorname{spt} S_j \subset \overline{B}_j$ for large enough j but $\underline{M}(S_j) \leq R$ is *not* clear so that $S_j \in \mathcal{C}_j$ is open.)

$$\underline{M}(S_j) \leq \underline{M}(T_j \llcorner B_{r_j}) + \underline{M}(\langle Q_j, r_j\rangle) + \underline{M}(\partial Y_j),$$

$$\underline{M}(\partial Y_j) = (\gamma_{m+1}^{-1}|\omega|^{-1}|Y_j(\omega)|)^{\frac{m}{m+1}} = (\gamma_{m+1}^{-1}|\omega|^{-1}|V_\omega(\partial Y_j, 0)|)^{\frac{m}{m+1}}$$

$$= (\gamma_{m+1}^{-1}|\omega|^{-1}|V_\omega(T_j^*, T_j)|)^{\frac{m}{m+1}}$$

$$\leq \text{(Opt. Isop. Theorem)} \ \underline{M}(T_j - T_j^*),$$

hence

$$\underline{M}(S_j) \leq \underline{M}(T_j \llcorner B_{r_j}) + \underline{M}(T_j \llcorner \complement B_{r_j} - \langle Q_j, r_j\rangle) + \underline{M}(\langle Q_j, r_j\rangle)$$

$$\leq \underline{M}(T_j \llcorner B_{r_j}) + \underline{M}(T_j \llcorner \complement B_{r_j}) + 2 \cdot \underline{M}(\langle Q_j, r_j\rangle)$$

Since we can choose r_j also to satisfy $\underline{M}(T_j \llcorner \partial B_{r_j}) = 0$ we end up with $\underline{M}(S_j) \leq \underline{M}(T_j) + \frac{2}{j}$ (so that we only have $\underline{M}(S_j) \leq R + \frac{2}{j}$ but not necessarily $\leq R$).

$$\mathcal{F}(S_j) = \underline{M}(S_j) + V_\omega(S_j, T_0)$$

$$= \underline{M}(T_j) + \underline{M}(S_j) - \underline{M}(T_j) + V_\omega(T_j, T_0)$$

$$+ V_\omega(S_j, T_0) - V_\omega(T_j, T_0)$$

$$\leq \mathcal{F}(T_j) + \frac{2}{j} + V_\omega(S_j, T_j),$$

$$V_\omega(S_j, T_j) = V_\omega(T_j^* + \partial Y_j, T_j) = V_\omega(T_j^*, T_j) + V_\omega(\partial Y_j, 0)$$

$$= 0 \quad \text{by the choice of } Y_j.$$

This proves the desired inequalities.

5. After passing to subsequences we have $S_j \to S$ for some $S \in \mathcal{R}_m$, $\partial S = \partial T_0$ and $\underline{M}(S) \leq R$ (which follows from $\underline{M}(S_j) \leq R + \frac{2}{j}$), hence $S \in \mathcal{C}_R$. Moreover, we know spt $S_j \subset \overline{B}_t$ for some ball of sufficiently large radius so that also spt $S \subset \overline{B}_t$. This implies

$$V_\omega(S_j, T_0) \xrightarrow[j \to \infty]{} V_\omega(S, T_0),$$

thus

$$\mathcal{F}(S) \leq \liminf_{j \to \infty} \mathcal{F}(S_j) \leq \liminf_{j \to \infty} \mathcal{F}(T_j).$$

If $U \in \mathcal{C}_R$ has compact support then $U \in \mathcal{C}_j$ for all $j \gg 1$, hence $\mathcal{F}(T_j) \leq \mathcal{F}(U)$ and we deduce $\mathcal{F}(S) \leq \mathcal{F}(U)$.

2.4 General Existence Theorems, Applications to Isoperimetric Problems

If $U \in \mathcal{C}_R$ has arbitrary support we then let U_s denote the projection of U onto the ball $\overline{B}_s$, i.e.

$$U_s := P_\# U, \quad P(y) = \begin{cases} y, & |y| \leq s \\ s \cdot \frac{y}{|y|}, & |y| \geq s \end{cases}$$

and obtain

$$\mathcal{F}(S) \leq \mathcal{F}(U_s).$$

The claim

$$\mathcal{F}(S) = \inf_{\mathcal{C}_R} \mathcal{F}$$

follows by passing to the limit $s \nearrow \infty$. □

The preceeding arguments, especially the "Key Lemma", also apply to the following *Isoperimetric Problem*.

Theorem 2.4.2 *Assume that $T_0 \in \mathcal{R}_m$ has compact support and finite mass. Let a constant mean curvature form $\Omega \neq 0$ and some number $c \in \mathbb{R}$ be given. Then the problem $\underline{M}(T) \to Min$ in the class $\mathcal{C} := \{T \in \mathcal{R}_m : \partial T = \partial T_0, V_\omega(T, T_0) = c\}$ admits a solution with compact support. Moreover, there is a real number μ such that the above solution T is a solution of the mean curvature equation with curvature form $\mu \cdot \Omega$ on $\mathbb{R}^{m+k} - \text{spt}(\partial T_0)$, i.e.*

$$\int_M (div_M X + \mu \cdot X \cdot \Omega(\vec{T})) \, d\mu_T = 0$$

for all $X \in C_0^1(\mathbb{R}^{m+k}, \mathbb{R}^{m+k})$, $\text{spt} X \cap \text{spt} \partial T_0 = \emptyset$. □

The above theorem generalizes known results for parametric surfaces: if one looks for a least area surface which together with a fixed reference surface encloses a prescribed (oriented) volume then the solution surface has constant mean curvature.

The *Proof of Theorem 2.4.2* is almost identical to that of Theorem 2.4.1:

1. $\mathcal{C}$ contains a current with compact support, e.g. $S_0 := T_0 + \partial Y_{t,\kappa}$ with $t > 0$, $\kappa \in \{-1, 1\}$ chosen to satisfy $V_\omega(\partial Y_{t,\kappa}, 0) = Y_{t,\kappa}(\omega) = c$. Thus we may consider the problems

$$\underline{M} \to \text{Min} \quad \text{in} \quad \mathcal{C}_j = \mathcal{C} \cap \{T \in \mathcal{R}_m : \text{spt} T \subset \overline{B}_j\}$$

at least for $j \gg 1$. Let T_j denote a solution (obtained by considering an $\underline{M}$-minimizing sequence U_ℓ in $\mathcal{C}_j$; we have $U_\ell \to U$ for some current U and since $\text{spt} U_\ell \cup \text{spt} U \subset \overline{B}_j$ it follows that $V_\omega(U_\ell, T_0) \to V_\omega(U, T_0)$ so that U is in $\mathcal{C}_j$).

2. With our previous notations we introduce the currents

$$T_j^* := T_j \,\llcorner\, B_{r_j} + \langle Q_j, r_j \rangle$$

(the "Key Lemma" can be shown to hold for the *mass* minimizers T_j and the corresponding currents Q_j) which are not necessarily in $\mathcal{C}_j$. In order to compensate the change of volume which results from cutting T_j and adding the slice we define the $(m+1)$-balls Y_j as before and obtain new currents

$$S_j := T_j^* + \partial Y_j$$

with uniformly bounded supports and the properties

$$\underline{M}(S_j) \leq \underline{M}(T_j) + \frac{2}{j}, \quad S_j \in \mathcal{C}_j \quad (\text{i.e. } V_\omega(S_j, T_j) = c).$$

3. A subsequence of $\{S_j\}$ then converges to a solution T of our problem.

4. The constant factor μ occurs as a Lagrange multiplier:
We may assume that there exists a field $Y \in C_0^1(\mathbb{R}^{m+k}, \mathbb{R}^{m+k})$ such that

$$\text{spt}\, Y \cap \text{spt}\, \partial T_0 = \emptyset \quad \text{and} \quad \int_M Y \cdot \Omega(\vec{T}) d\mu_T \neq 0$$

for if the above integral vanishes for all such Y then it is easy to check that $\Omega(\vec{T}) = 0$, μ_T – a.e. and

$$\int_M \text{div}_M X \, d\mu_T = 0,$$

$X \in C_0^1(\mathbb{R}^{m+k}, \mathbb{R}^{m+k})$, $\text{spt}\, X \cap \text{spt}\, \partial T_0 = \emptyset$, holds. For X as above let

$$\varphi(s, t, z) := z + s \cdot X(z) + t \cdot Y(z)$$

and define

$$T_{s,t} = \varphi(s, t, \cdot)_\# T.$$

Then $\partial T_{s,t} = \partial T_0$, $g(s,t) := V_\omega(T_{s,t}, T_0) = V_\omega(T_{s,t}, T) + c$ so that $\frac{\partial}{\partial t} g(0,0) = \int_M Y \cdot \Omega(\vec{T}) \, d\mu_T \neq 0$. Observing $g(0,0) = c$ we fix a curve $\sigma(s)$ such that $\sigma(0) = 0$ and $g(s, \sigma(s)) \equiv c$. This implies

$$T_{s,\sigma(s)} = \varphi(s, \sigma(s), \cdot)_\# T \in \mathcal{C},$$

hence

2.4 General Existence Theorems, Applications to Isoperimetric Problems

$$0 = \frac{d}{ds/0} \underline{M}(T_{s,\sigma(s)})$$

$$= \int_M \operatorname{div}_M \left(\frac{\partial}{\partial s/0} \varphi(s,\sigma(s))\right) d\mu_T$$

$$= \int_M \operatorname{div}_M X \cdot d\mu_T + \sigma'(0) \int_M \operatorname{div}_M Y \, d\mu_T.$$

In view of

$$0 = \frac{d}{ds/0} g(s,\sigma(s)) = \frac{\partial}{\partial s} g(0,0) + \sigma'(0) \frac{\partial}{\partial t} g(0,0)$$

$$= \int_M X \cdot \Omega(\vec{T}) d\mu_T + \sigma'(0) \int_M Y \cdot \Omega(\vec{T}) d\mu_T$$

we arrive at

$$0 = \int_M \operatorname{div}_M X \cdot d\mu_T + \left(\frac{-\int_M \operatorname{div}_M Y \, d\mu_T}{\int_M Y \cdot \Omega(\vec{T}) d\mu_T}\right) \cdot \int_M X \cdot \Omega(\vec{T}) \, d\mu_T$$

so that

$$\mu = \frac{-\int_M \operatorname{div}_M Y \cdot d\mu_T}{\int_M Y \cdot \Omega(\vec{T}) \, d\mu_T}.$$

□

Since T solves an equation of mean curvature type we get as

Corollary: *Allard's regularity theorem applies to the solution of the isoperimetric problem defined in Theorem 2.4.2.*

We now want to extend Theorem 2.4.1 to the case of nonconstant mean curvature forms Ω. Clearly we have to modify our approach since for variable curvatures Ω the attaching of "suitable" m-spheres ∂Y which compensate the loss of volume is now impossible. On the other hand our main purpose is to solve the M.C.P. and we already know that this is only possible under a smallness condition of the form (2.4.1) relating $|\Omega|$ and the minimal area spanned by the boundary current $\Gamma \in \mathcal{R}_{m+1}$. It is therefore natural to ask if curvature conditions of type (2.4.1) are also sufficient for proving the existence of $\mathcal{F}$-minimizing currents. The following result is taken from [17].

Theorem 2.4.3 *Let Ω denote a bounded continuous mean curvature form $\mathbb{R}^{m+k} \times \Lambda_m \mathbb{R}^{m+k} \to \mathbb{R}^{m+k}$ whose associated scalar mean curvature form ω is weakly closed and suppose that we are given an $(m-1)$-current $\Gamma \in \mathcal{R}_{m-1}$ such that $\mathrm{spt}\,\Gamma$ is compact and $\underline{M}(\Gamma) < \infty$, $\partial \Gamma = 0$. Let $\Sigma_0 \in \mathcal{R}_m$ denote a mass minimizer for the boundary Γ, i.e. $\underline{M}(\Sigma_0) = \inf\{\underline{M}(S) : S \in \mathcal{R}_m,\ \partial S = \Gamma\}$. Define*

$$R := (2m+1) \cdot \underline{M}(\Sigma_0)$$

and impose the curvature bound

$$|\Omega| = |\omega| \leq \gamma_{m+1}^{-1} \cdot 2^{-1/m} \cdot (m+1)^{-1/m} \cdot \underline{M}(\Sigma_0)^{-1/m} \quad (2.4.2)$$
$$:= c_m \cdot \underline{M}(\Sigma_0)^{-1/m}.$$

Then the problem

$$\mathcal{F}(T) := \underline{M}(T) + V_\omega(T, \Sigma_0) \to \min$$

$$in \ \ \mathcal{C}_R := \{S \in \mathcal{R}_m : \partial S = \Gamma,\ \underline{M}(S) \leq R\}$$

has a solution with compact support. If (2.4.2) is replaced by the stronger condition (2.4.1)

$$|\Omega| < \frac{m}{m+1} \cdot c_m \underline{M}(\Sigma_0)^{-1/m}$$

then the above minimzer is a solution of the M.C.P.

Remarks:

1. If T_0 is any finite mass current with compact support and boundary Γ then the statement of Theorem 2.4.3 clearly holds for $\mathcal{F}'(T) := \underline{M}(T) + V_\omega(T, T_0)$ for the reason that $\mathcal{F}$ and $\mathcal{F}'$ only differ by an additive constant.

2. In case $m = 2$, $k = 1$ and $\Gamma = S^1 \times \{0\}$ condition (2.4.2) reads as

$$|\Omega| \leq \sqrt{6} \approx 2.45$$

   so that in general one can not expect (2.4.2) to be sufficient for our minimizer to be a solution of the M.C.P. According to Heinz we can expect this only for $|\Omega| \leq 2$, and we know that the minimizer solves the M.C.P. if $|\Omega| < 2 \cdot \frac{\sqrt{2}}{\sqrt{3}} \approx 1.633$.

2.4 General Existence Theorems, Applications to Isoperimetric Problems

Proof of Theorem 2.4.3: We proceed similar as in Theorem 2.4.1 and define

$$\mathcal{C}_j := \mathcal{C}_R \cap \{S \in \mathcal{R}_m : \operatorname{spt} S \subset \overline{B}_j\}$$

observing $\Sigma_0 \in \mathcal{C}_j$ at least for $j \in \mathbb{N}$ large enough. Let T_j denote a current with the property

$$\mathcal{F}(T_j) = \inf_{\mathcal{C}_j} \mathcal{F}$$

and consider a mass minimizer Q_j for the boundary $T_j - \Sigma_0$. Without proof we then observe that the Key Lemma is still true which provides us with a sequence of suitable slicing radii r_j for which $\underline{M}(\langle Q_j, r_j \rangle) \leq j^{-1}$ (and in addition $\underline{M}(Q_j \llcorner \partial B_{r_j}) = 0 = \underline{M}(T_j \llcorner \partial B_{r_j})$). Now let $T_j^* := T_j \llcorner B_{r_j} + \langle Q_j, r_j \rangle$ so that

$$\operatorname{spt} T_j^* \subset \overline{B}_j, \quad \underline{M}(T_j^*) \leq R + j^{-1} \quad \text{and} \quad \partial T_j^* = \partial \Sigma_0.$$

Moreover

$$\begin{aligned}
\mathcal{F}(T_j^*) &= \underline{M}(T_j \llcorner B_{r_j} + \langle Q_j, r_j \rangle) + V_\omega(T_j^*, \Sigma_0) \\
&\leq \underline{M}(T_j \llcorner B_{r_j}) + j^{-1} + V_\omega(T_j^*, T_j) + V_\omega(T_j, \Sigma_0)
\end{aligned}$$

and from the Optimal Isoperimetric Theorem we infer

$$\begin{aligned}
V_\omega(T_j^*, T_j) &\leq \gamma_{m+1} \cdot |\omega| \cdot \underline{M}(T_j^* - T_j)^{1+1/m} \\
&\leq \gamma_{m+1} \cdot |\omega| \cdot (\underline{M}(T_j) + j^{-1})^{1/m} \cdot \underline{M}(T_j \llcorner \complement B_{r_j} - \langle Q_j, r_j \rangle) \\
&\leq \gamma_{m+1} \cdot |\omega| \cdot (R + j^{-1})^{1/m} \cdot (\underline{M}(T_j \llcorner \complement B_{r_j}) + j^{-1})
\end{aligned}$$

so that

$$\begin{aligned}
\mathcal{F}(T_j^*) &\leq \underline{M}(T_j) + V_\omega(T_j, \Sigma_0) + j^{-1} \left\{ 1 + |\omega| \cdot \gamma_{m+1} (R + j^{-1})^{1/m} \right\} \\
&\quad + [\gamma_{m+1} \cdot |\omega| \, (R + j^{-1})^{1/m} - 1] \cdot \underline{M}(T_j \llcorner \complement B_{r_j}).
\end{aligned}$$

From the curvature bound (2.4.2) we deduce

$$\begin{aligned}
\gamma_{m+1} \cdot |\omega| \cdot (R + j^{-1})^{1/m} - 1 &\leq 2^{-1/m} \, (m+1)^{-1/m} \, (2m+1)^{1/m} \, R^{-1/m} \, (R + j^{-1})^{1/m} - 1 \\
&= \left(\frac{2m+1}{2m+2}\right)^{1/m} \left(\frac{R + j^{-1}}{R}\right)^{1/m} - 1 \leq 0
\end{aligned}$$

for $j \gg 1$, hence

$$\begin{cases} \mathcal{F}(T_j^*) \leq \mathcal{F}(T_j) + c \cdot j^{-1}, \\ c := 1 + |\omega| \cdot \gamma_{m+1} \, (R+1)^{1/m}. \end{cases} \quad (2.4.3)$$

As before (recall $\underline{M}(T_j^*) \leq R + \frac{1}{j}$ and $\sup_j(\operatorname{diam}\operatorname{spt} T_j^*) < \infty$) we find $T \in \mathcal{C}_R$ such that $T_j^* \to T$ and $V_\omega(T_j^*, \Sigma_0) \to V_\omega(T, \Sigma_0)$. From (2.4.3) it is easy to deduce that T minimizes among currents U in $\mathcal{C}_R$ with compact support since then $\mathcal{F}(T_j) \leq \mathcal{F}(U)$. If $U \in \mathcal{C}_R$ has arbitrary support we let U_r denote the projection onto a large ball $\overline{B}_r$ and get $\mathcal{F}(T) \leq \mathcal{F}(U_r) = \underline{M}(U_r) + V_\omega(U_r, \Sigma_0) = \mathcal{F}(U) + V_\omega(U_r, U)$,

$$|V_\omega(U_r, U)| \leq \gamma_{m+1}|\omega| \cdot \underline{M}(U - U_r)^{1+1/m} \xrightarrow[r \to \infty]{} 0$$

which shows that T is minimizing in $\mathcal{C}_R$. □

Concerning the problem of minimizing mass subject to a volume constraint the reader is also referred to the very interesting recent paper of Duzaar and Steffen [20] which contains a detailed analysis of the codimension one case.

2.5 Tangent Cones, Small Solutions, Closed Hypersurfaces

The study of oriented tangent cones is a powerfull tool for describing the regularity properties of mean curvature currents since the structure of a tangent cone at $x_0 \in \operatorname{spt} T$ infinitesimaly reflects the behaviour of $\operatorname{spt} T$ near x_0. We can prove the following result.

Theorem 2.5.1 *Assume that Ω is a given mean curvature form whose associated scalar mean curvature form is bounded, continuous and weakly closed. Suppose further that $\Gamma \in \mathcal{R}_{m-1}$ has finite mass and no boundary. Consider a class $\mathcal{C} \subset \mathcal{R}_m$ of currents T satisfying*

$$\partial T = \Gamma, \quad \underline{M}(T) \leq R$$

and where also a side condition of the form $\operatorname{spt} T \subset K$ with K denoting the closure of a smooth bounded region may be included. Fix some reference current $T_0 \in \mathcal{C}$ and suppose that there exists $T \in \mathcal{C}$ such that

$$\mathcal{F}(T) := \underline{M}(T) + V_\omega(T, T_0) = \inf_{\mathcal{C}_R} \mathcal{F}, \quad \underline{M}(T) < R.$$

Then the following results are true: Let $X_0 \in \operatorname{spt} T - \operatorname{spt} \Gamma$ and a sequence $\{\lambda_i\} \subset \mathbb{R}$, $\lambda_i \downarrow 0$, be given. Then there is a subsequence $\{\lambda_i^*\} \subset \{\lambda_i\}$ such that

$$T_i := \left(\eta_{x_0, \lambda_i^*}\right)_\# T \to: C \quad \text{in } \mathcal{D}_m,$$

$$\eta_{x_0, \lambda_i^*}(z) := \lambda_i^{*-1}(z - x_0),$$

where C is integer multiplicity. We have:

2.5 Tangent Cones, Small Solutions, Closed Hypersurfaces

(i) $\partial C = 0$, $\Theta^m(\mu_T, x_0) := \lim_{\rho \downarrow 0} \alpha_m^{-1} \cdot \rho^{-m} \mu_T(B_\rho(x_0)) = \Theta^m(\mu_C, 0)$ and $\mu_{T_i} \to \mu_C$ in the sense of Radon measures.

(ii) C is a cone, i.e. $(\eta_{0,\lambda})_\# C = C$ for all $\lambda > 0$.

(iii) C is locally mass minimizing, i.e. we have
$$\underline{M}_W(C) \leq \underline{M}_W(C + S)$$
for all open sets $W \subset \mathbb{R}^{m+k}$ with compact closure and all $S \in \mathcal{R}_m$, spt $S \subset W$, $\partial S = 0$.

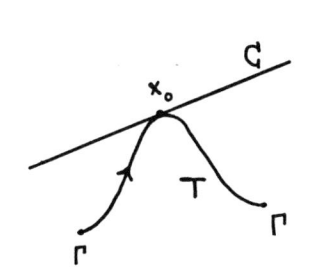

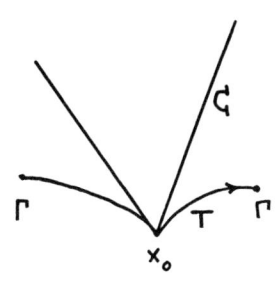

regular case: C is a plane   singular case: C is a cone.

Remarks:

1. Similar results are true if T is a solution of the isoperimetric problem of Theorem 2.4.2.

2. In sections 2.3, 2.4 we gave sufficient conditions for the existence of $\mathcal{F}$-minimizers satisfying in addition the strict inequality $\underline{M}(T) < R$ for suitable R and under appropriate curvature bounds. The property $\underline{M}(T) < R$ is needed for proving an Euler equation of the form
$$\int_M (\mathrm{div}_M X + X \cdot \Omega(\vec{T})) \, d\mu_T = \int_{M \cap \partial K} X \cdot F \, d\mu_T,$$
$$X \in C_0^1(\mathbb{R}^{m+k}, \mathbb{R}^{m+k}), \text{ spt } X \cap \Gamma = \emptyset,$$
with right-hand-side $F \in L^\infty(\mathbb{R}^{m+k}, \mathbb{R}^{m+k}; \mu_T)$ resulting from the possible contact of spt T with ∂K if we work in a class of currents being contained in K. If we drop this condition then the above equation reduces to the mean curvature equation.

3. Part i) of Theorem 2.5.1 especially includes the fact that the m–dimensional density exists at every point $x_0 \in \operatorname{spt} T - \operatorname{spt} \Gamma$ which is not necessarily true for an arbitrary $T \in \mathcal{C}$. The existence of $\Theta^m(\mu_T, x_0)$ is a consequence of the *monotonicity formula* which follows from the above Euler equation. Here we do not exploit the $\mathcal{F}$-minimality of T but only use the fact that the varifold associated to T is of bounded first variation, we refer to [74, 17.8].

The minimizing property of T is necessary for proving that the limit C is area minimizing. The existence of the limit C follows from the local mass bound

$$\underline{M}_{B_r}(T_i) = \alpha_m r^m \cdot \mu_T\bigl(B_{\lambda_i r}(x_0)\bigr) \,/\, \alpha_m \cdot (\lambda_i r)^m$$

observing that the right-hand-side has a limit as $i \to \infty$.

4. During the proof of Theorem 2.5.1 one possibly has to handle the case $x_0 \in (\operatorname{spt} T - \operatorname{spt} \Gamma) \cap \partial K$ which causes some technical difficulties. We refer to [10] or [15] for details. As in the case of minimal surfaces the *question of uniqueness of oriented tangent cones is completely open.*

5. Similar to [74, §37] we can apply Theorem 2.5.1 combined with a dimension reduction argument to prove Theorem 2.3.7.

We now want to describe in some detail how to obtain the existence of mass minimizing tangent cones for solutions of the isoperimetric problem.

Suppose that the assumptions of Theorem 2.4.2 are satisfied. We may assume that $x_0 = 0 \in \operatorname{spt} T - \operatorname{spt} \partial T$ where T now denotes a solution of

$$\underline{M}(T) \to \operatorname{Min} \text{ in } \mathcal{C} = \{ S \in \mathcal{R}_m : \partial T = \partial T_0,\ V_\omega(T, T_0) = c \}.$$

Let $T_j := (\eta_{\lambda_j})_\# T$ with $\eta_t(x) := \frac{1}{t}x$. Since T solves the mean curvature equation with curvature form $\mu \cdot \Omega$ (for a suitable real number μ) we may apply [74, 17.8] to get

$$\sup_{j \in \mathbb{N}} \underline{M}_{B_r}(T_j) < \infty$$

for any $r > 0$. Thus (after passing to a subsequence) we see $T_j \to C$ for an integer multiplicity current C with $\partial C = 0$.

We want to show that C is locally mass minimizing: fix a compact set $L \subset \mathbb{R}^{m+k}$ and a smooth function $\phi : \mathbb{R}^{m+k} \to [0,1]$ with compact support and with the property $\phi = 1$ in a neighborhood of L. For $0 \leq t \leq 1$ we let $W_t := \{ z \in \mathbb{R}^{m+k} : \phi(z) > t \}$. Quoting [74, 31.2] we may write

2.5 Tangent Cones, Small Solutions, Closed Hypersurfaces

$$C - T_j = \partial R_j + S_j,$$
$$\underline{M}_{W_0}(R_j) + \underline{M}_{W_0}(S_j) \to 0 \quad (2.5.1)$$

for suitable integer multiplicity currents R_j and S_j. By elementary slicing theory we can find $0 < \alpha < 1$ and currents P_j such that

$$\partial(R_j \llcorner W_j) = (\partial R_j) \llcorner W_\alpha + P_j \quad (2.5.2)$$

and with

$$\underline{M}(P_j) \to 0, \quad \mathrm{spt}\, P_j \subset \partial W_\alpha.$$

In addition we may also assume

$$\underline{M}(T_j \llcorner \partial W_\alpha) = 0, \quad \underline{M}(C \llcorner \partial W_\alpha) = 0. \quad (2.5.3)$$

From (2.5.1) and (2.5.2) we deduce

$$C \llcorner W_\alpha = T_j \llcorner W_\alpha + \partial \tilde{R}_j + \tilde{S}_j \quad (2.5.4)$$

where

$$\tilde{R}_j := R_j \llcorner W_\alpha, \quad \tilde{S}_j := S_j \llcorner W_\alpha - P_j,$$

and

$$\underline{M}(\tilde{R}_j) + \underline{M}(\tilde{S}_j) \to 0. \quad (2.5.5)$$

Next let X denote a given m–current satisfying $\partial X = 0$ and $\mathrm{spt}\, X \subset L$. Abbreviating $Z_j := X + \partial \tilde{R}_j$ we have $\mathrm{spt}\, Z_j \subset \overline{W}_\alpha$ and (recall (2.5.4))

$$\begin{aligned} \underline{M}_{W_\alpha}(C + X) &= \underline{M}_{W_\alpha}(T_j + X + \partial \tilde{R}_j + \tilde{S}_j) \\ &\geq \underline{M}_{W_\alpha}(T_j + Z_j) - \underline{M}(\tilde{S}_j). \end{aligned} \quad (2.5.6)$$

For arbitrary $t < \alpha$ we find ($W_t^j := \lambda_j W_t$)

$$\underline{M}_{W_t}(T_j + Z_j) = \lambda_j^{-m} \underline{M}_{W_t^j}\left(T + (\eta_{\lambda_j^{-1}})_\# Z_j\right)$$

and

$$\underline{M}_{W_t^j}\left(T + (\eta_{\lambda_j^{-1}})_\# Z_j\right) - \underline{M}_{W_t^j}(T) = \underline{M}\left(T + (\eta_{\lambda_j^{-1}})_\# Z_j\right) - \underline{M}(T).$$

Suppose now that there exists a sequence of positive numbers $\varepsilon_j \downarrow 0$ with the property

$$\underline{M}\left(T + (\eta_{\lambda_j^{-1}})_\# Z_j\right) - \underline{M}(T) \geq -\varepsilon_j \lambda_j^m. \quad (2.5.7)$$

In this case we deduce
$$\underline{\underline{M}}_{W_t}(T_j + Z_j) - \underline{\underline{M}}_{W_t}(T_j) \geq -\varepsilon_j$$
which gives after passing to the limit $t \uparrow \alpha$
$$\underline{\underline{M}}_{W_\alpha}(T_j + Z_j) - \underline{\underline{M}}_{W_\alpha}(T_j) \geq -\varepsilon_j - \underline{\underline{M}}(P_j). \qquad (2.5.8)$$
Here we have used (recall (2.5.3))
$$\lim_{t \uparrow \alpha} \underline{\underline{M}}_{W_t}(T_j) = \mu_{T_j}(\overline{W}_\alpha) = \underline{\underline{M}}_{W_\alpha}(T_j)$$
and
$$\lim_{t \uparrow \alpha} \underline{\underline{M}}_{W_t}(T_j + Z_j)$$
$$= \mu_{T_j+Z_j}(\overline{W}_\alpha) = \mu_{T_j+Z_j}(W_\alpha) + \mu_{T_j+X+\partial \tilde{R}_j}(\partial W_\alpha)$$
$$= \underline{\underline{M}}_{W_\alpha}(T_j + Z_j) + \underline{\underline{M}}(P_j).$$

Remembering the decomposition (2.5.4) estimate (2.5.8) implies
$$\underline{\underline{M}}_{W_\alpha}(C + X) \geq \underline{\underline{M}}_{W_\alpha}(T_j) - \varepsilon_j - \underline{\underline{M}}(P_j) - \underline{\underline{M}}(S_j),$$
therefore by (2.5.5) and the lower semicontinuity of the mass
$$\underline{\underline{M}}_{W_\alpha}(C + X) \geq \underline{\underline{M}}_{W_\alpha}(C)$$
which proves the local minimality of C. It remains to establish (2.5.7). This will be done under the following additional assumption imposed on T: suppose that
$$\mu_T\Big(\{z \in \mathbb{R}^{m+k} - \mathrm{spt}\,\partial T : |\Omega(\vec{T}(z))| > 0\}\Big) > 0. \qquad (2.5.9)$$

(We will comment (2.5.9) later on.) In this case there exists a vectorfield $Y \in C_0^1(\mathbb{R}^{m+k}, \mathbb{R}^{m+k})$, $\mathrm{spt}\,Y \cap \mathrm{spt}\,\partial T = \emptyset$, such that
$$\alpha := \int Y \cdot \Omega(\vec{T}) d\mu_T > 0.$$

Let $U_s := \phi(s, \cdot)_\# T$, $\phi(s, z) := z + sY(z)$. We want to calculate s_j such that
$$V_\omega(U_{s_j} + \hat{Z}_j, T_0) = c \qquad (2.5.10)$$
is valid for $\hat{Z}_j := (\eta_{\lambda_j^{-1}})_\# Z_j$. Clearly (2.5.10) is equivalent to

2.5 Tangent Cones, Small Solutions, Closed Hypersurfaces

$$V_\omega(U_{s_j}, T) = V_\omega(-\hat{Z}_j, 0) =: \beta_j.$$

The function $f(s) := V_\omega(U_s, T)$ satisfies $f(0) = 0$, $f'(0) = \alpha > 0$, especially we find $\delta > 0$ such that $f' \geq \frac{\alpha}{2}$ on $[-\delta, \delta]$ which implies

$$f([-\delta, \delta]) \supset \left[-\frac{\alpha}{2}\delta, \frac{\alpha}{2}\delta\right].$$

On the other hand it is easy to check the bound

$$|\beta_j| \leq A \lambda_j^{m+1}$$

for some absolute constant $A > 0$ so that we may assume

$$\beta_j \in \left[-\frac{\alpha}{2}\delta, \frac{\alpha}{2}\delta\right]$$

for all $j \in \mathbb{N}$. Hence there exists a unique number $s_j \in [-\delta, \delta]$ with $f(s_j) = \beta_j$ and one easily proves that

$$|s_j| \leq \frac{2}{\alpha} A \lambda_j^{m+1}. \tag{2.5.11}$$

Let $U_j := U_{s_j}$. Recalling (2.5.10) the minimality of T implies

$$\underline{M}(T) \leq \underline{M}(U_j + \hat{Z}_j).$$

On the other hand one can arrange

$$\operatorname{spt} \hat{Z}_j \cap \operatorname{spt} Y = \emptyset$$

so that

$$\underline{M}(T + \hat{Z}_j) - \underline{M}(U_j + \hat{Z}_j) = \underline{M}(T) - \underline{M}(U_j).$$

Collecting our results we arrive at

$$\underline{M}(T + \hat{Z}_j) - \underline{M}(T) \geq \underline{M}(T + \hat{Z}_j) - \underline{M}(U_j + \hat{Z}_j)$$

$$= \underline{M}(T) - \underline{M}(U_j) = \frac{1}{s_j}[\underline{M}(T) - \underline{M}(\phi(s_j, \cdot)_\# T)] s_j$$

and since

$$\frac{1}{s_j}[\ldots] \xrightarrow[j \to \infty]{} -\int_M \operatorname{div}_M Y \, d\mu_T$$

we get

$$\underline{M}(T + \hat{Z}_j) - \underline{M}(T) \geq -B|s_j|$$

for some $B > 0$ and all $j \gg 1$. Recalling (2.5.11) we have demonstrated (2.5.7). That C is a cone follows along the lines of [74].

It remains to discuss condition (2.5.9). Suppose that the boundary ∂T_0 is of the form $\partial T_0 = \Theta_0 [\![\Sigma]\!]$ for some $\Theta_0 \in \mathbb{N}$ and a compact, oriented $(m-1)$-dimensional submanifold $\Sigma \subset \mathbb{R}^{m+k}$ without boundary. Then we showed in [15] that $0 \times \partial T_0 \not\subset C$ is sufficient for (2.5.9).

So our nondegeneracy assumption holds if T_0 and the cone $0 \times \partial T_0$ over ∂T_0 enclose a volume different from c.

In sections 2.2 and 2.3 we already discussed the advantage of introducing the oriented curvature volume V_ω whose analysis leads to rather general existence theorems for the M.C.P. On the other hand if α is an m–form such that $\omega = d\alpha$ we have

$$\mathcal{F}(T) = \underline{M}(T) + T(\alpha)$$

up to an additive constant and we now want to discuss conditions on ω and the class of admissible currents under which it is very easy to obtain solutions of the M.C.P. corresponding to the so-called small solutions in the two-dimensional parametric setting.

Theorem 2.5.2 *Suppose that Ω is a mean curvature form of class C^1 such that $\sup_{|z|\leq 1}|\Omega_z| < m+1$ and define*

$$\langle \alpha(z), \xi_1 \wedge \ldots \wedge \xi_m \rangle := \int_0^1 t^m \cdot \langle \omega(t \cdot z), z \wedge \xi_1 \wedge \ldots \wedge \xi_m \rangle \, dt$$

for $z, \xi_1, \ldots, \xi_m \in \mathbb{R}^{m+k}$. Then $d\alpha = \omega$ and the problem

$$\mathcal{G}(T) := \underline{M}(T) + T(\alpha) \to \text{Min}$$

in $\mathcal{C} := \{T \in \mathcal{R}_m : \operatorname{spt} T \subset \overline{B}_1, \partial T = \Gamma\}$ has a solution for any $\Gamma \in \mathcal{R}_{m-1}$, $\partial \Gamma = 0$, $\operatorname{spt} \Gamma \subset \overline{B}_1$.

Proof: One easily checks $\sup_{|z|\leq 1}|\alpha_z| \leq 1-\varepsilon$ for some $\varepsilon > 0$ so that

$$\mathcal{G}(T) \geq \varepsilon \cdot \underline{M}(T)$$

for $T \in \mathcal{C}$. This implies compactness of $\mathcal{G}$-minimizing sequences. $\square$

Theorem 2.5.3 *Suppose in addition to the hypothesis of Theorem 2.5.2 that $\operatorname{spt} \Gamma$ has positive distance to ∂B and that*

$$|\Omega_z| < m, \quad |z| = 1.$$

2.5 Tangent Cones, Small Solutions, Closed Hypersurfaces

Then the solution defined in Theorem 2.5.2 has positive distance to ∂B and solves

$$\int_M \left(div_M X + X \cdot \Omega(\vec{T}) \right) d\mu_T = 0$$

for all $X \in C_0^1(\mathbb{R}^{m+k}, \mathbb{R}^{m+k})$, $\operatorname{spt} X \cap \operatorname{spt} \Gamma = \emptyset$.

Proof: For small positive t let

$$P(t, z) := \begin{cases} z, & |z| \leq 1 - t \\ (1-t)z/|z|, & |z| \geq 1 - t; \end{cases}$$

then $P(t, \cdot)_\# T \in \mathcal{C}$ and by calculating $\mathcal{G}(P(t, \cdot)_\# T)$ we arrive at the contradiction $\mathcal{G}(P(t, \cdot)_\# T) < \mathcal{G}(T)$ if $\operatorname{spt} T$ touches ∂B. Hence we get

$$0 = \frac{d}{ds/0} \mathcal{G}(\phi(s, \cdot)_\# T)$$

for $\phi(s, z) := z + s \cdot X(z)$, $X \in C_0^1(\mathbb{R}^{m+k}, \mathbb{R}^{m+k})$, $\operatorname{spt} X \cap \operatorname{spt} \Gamma = \emptyset$. This is the desired Euler equation. □

We close this section with a discussion of the following problem in the codimension 1 case:

> Given a smooth bounded region $\Omega \subset \mathbb{R}^{m+1}$ compactly contained in some other domain G (not necessarily bounded) and a function $h : \overline{G} \to \mathbb{R}$ we want to prove the existence of a set $A \subset \mathbb{R}^{m+1}$ such that $\overline{\Omega} \subset A \subset \overline{G}$ holds and the additional property that the mean curvature $H_{\partial A}$ of ∂A calculated w.r.t. the interior normal ν_A is given by h.

Thus we want to enclose Ω by a body A whose curvature at each point of ∂A is just h. The right setting for solving this problem is provided by the so-called *Caccioppoli sets* or sets of finite perimeter.

Definition: An $\mathcal{L}^{m+1}$-measurable set A has locally finite perimeter iff $\int_U |\nabla \mathbf{1}_A| < \infty$ for any U open in $\mathbb{R}^{m+k}$ with compact closure. Here

$$\mathbf{1}_A(x) = \begin{cases} 1, & x \in A \\ 0, & \text{else} \end{cases}$$

Associated to a set A of finite perimeter is the *reduced boundary* $\partial^* A$ on which we can define the "inward pointing unit normal ν_A" (see [48]). In the language of currents we have

$$\partial \llbracket A \rrbracket = \underline{\tau}(\partial^* A, \nu_A, 1) \in \mathcal{R}_m$$

where $[\![A]\!] \in \mathcal{R}_{m+1}$ is the current "integration over the set A". It is easy to check that

$$\underline{\underline{M}}_U(\partial[\![A]\!]) = \int_U |\nabla \mathbf{1}_A|,$$

$$\underline{\underline{M}}([\![A]\!]) + \underline{\underline{M}}(\partial[\![A]\!]) = \mathcal{L}^{m+1}(A) + \int |\nabla \mathbf{1}_A|,$$

and the isoperimetric inequality now reads

$$\underline{\underline{M}}([\![A]\!]) \le \gamma_{m+1} \underline{\underline{M}}(\partial[\![A]\!])^{\frac{m+1}{m}}.$$

Definition: Let A denote a Cacciopploli set. We say that the boundary of A has prescribed mean curvature h with respect to ν_A iff

$$\int_{\partial^* A} (\operatorname{div}_{\partial^* A} X + h \nu_A \cdot X) \, d\mathcal{H}^m = 0$$

for all $X \in C_0^1(\mathbb{R}^{m+1}, \mathbb{R}^{m+1})$.

For sets A with smooth boundary this is equivalent to

$$\underline{\underline{H}}_{\partial A}(z) = h(z) \nu_A(z), \quad z \in \partial A,$$

where $\underline{\underline{H}}_{\partial A}$ is the mean curvature vector of ∂A in the sense of differential geometry.

Next we observe that

$$\frac{d}{dt/0} \mathcal{F}(f_t(A)) = \int_{\partial^* A} (\operatorname{div}_{\partial^* A} X + X h \nu_A) \, d\mathcal{H}^m$$

whenever f_t is a family of smooth diffeomorphisms such that $f_0 = \operatorname{Id}$ and $f_t - \operatorname{Id} = 0$ outside a compact set $K \subset \mathbb{R}^{m+1}$. Here $X := \frac{\partial}{\partial t} f_t$ and

$$\mathcal{F}(A) := \underline{\underline{M}}(\partial[\![A]\!]) - \int_A h \cdot d\mathcal{L}^{m+1}.$$

For the proof one observes

$$\frac{d}{dt/0} \mathcal{F}(f_t(A)) = \frac{d}{dt/0} \underline{\underline{M}}((f_t)_\# \partial[\![A]\!]) - \frac{d}{dt/0} \int_{f_t(A)} h \, d\mathcal{L}^{m+1}$$

$$= \int_{\partial^* A} \operatorname{div}_{\partial^* A} X \, d\mathcal{H}^m - \frac{d}{dt/0} \int_A h(f_t) \cdot \det Df_t \, d\mathcal{L}^{m+1}$$

and

2.5 Tangent Cones, Small Solutions, Closed Hypersurfaces

$$\frac{d}{dt/0}\int_A h(f_t)\det Df_t\, d\mathcal{L}^{m+1}$$

$$= \int_A (\nabla h \cdot X + h \cdot \operatorname{div} X)\, d\mathcal{L}^{m+1}$$

$$= \int_A \operatorname{div}(h \cdot X)\, d\mathcal{L}^{m+1}$$

$$= -\int_{\partial^* A} h\, \nu_A \cdot X\, d\mathcal{H}^m$$

by a version of the Gauss theorem for Caccioppoli sets.

Suggested by the above discussion we define

$$\mathcal{C} := \{A \subset \mathbb{R}^{m+1} : A \text{ is a Caccioppoli set satisfying } \overline{\Omega} \subset A \subset \overline{G}$$
$$\text{and } \underline{M}([\![A]\!]) + \underline{M}(\partial[\![A]\!]) < \infty\}$$

and $\mathcal{F} : \mathcal{C} \to \mathbb{R}$, $\mathcal{F}(A) = \underline{M}(\partial[\![A]\!]) - \int_A h\, d\mathcal{L}^{m+1}$.

Theorem 2.5.4 *Suppose that $h : \overline{G} \to \mathbb{R}$ is continous. Then the problem $\mathcal{F} \to \min$ in $\mathcal{C}$ has a solution if one of the following conditions holds:*

(i) $\overline{G}$ *compact* or

(ii) $h \in L^1(G)$ or

(iii) $\|h\|_{L^{m+1}(G)} \cdot \gamma_{m+1}^{m/(m+1)} < 1$.

Let $H_{\partial G}$ ($H_{\partial \Omega}$) denote the scalar mean curvature of ∂G ($\partial \Omega$) calculated w.r.t. the inward pointing unit normal. If

$$h(x) \leq H_{\partial G}(x), \quad x \in \partial G, \quad \text{and}$$

$$h(x) \geq H_{\partial \Omega}(x), \quad x \in \partial \Omega,$$

then the above minimizer A satisfies

$$\int_{\partial^* A} (\operatorname{div}_{\partial^* A} X + h \cdot \nu_A \cdot X)\, d\mathcal{H}^m = 0$$

for all $X \in C_0^1(\mathbb{R}^{m+1}, \mathbb{R}^{m+1})$. In case $m \leq 6$ A is a classical (i.e. smooth) solution.

Proof: We have the bounds

$$\mathcal{F}(A) \geq \underline{M}(\partial[\![A]\!]) - \begin{cases} \sup_{G}|h| \cdot \mathcal{L}^{m+1}(G) & \text{(in case i))} \\ \|h\|_{L^1(G)} & \text{(in case ii))} \\ \underline{M}(\partial[\![A]\!]) \cdot \gamma_{m+1}^{\frac{m}{m+1}} \|h\|_{L^{m+1}(G)} & \text{(in case iii))} \end{cases}$$

so that $\inf_{\mathcal{C}} \mathcal{F} > -\infty$. Consider an $\mathcal{F}$-minimizing sequence A_n. From the above estimates we get $\sup_{n \in \mathbb{N}} \underline{M}(\partial[\![A_n]\!]) < \infty$ so that $\mathcal{L}^{m+1}(A_n)$ is also bounded by the isoperimetric inequality. By BV–compactness there is some $A \in \mathcal{C}$ such that

$$\mathbf{1}_{A_n} \to \mathbf{1}_A \quad \text{in } L^1_{\text{loc}} \text{ and a.e.,}$$

$$\underline{M}(\partial[\![A]\!]) \leq \liminf_{h \to \infty} \underline{M}(\partial[\![A_n]\!]).$$

It remains to show $\mathcal{F}(A) \leq \liminf_{n \to \infty} \mathcal{F}(A_n)$ and we consider for example case iii): for a fixed radius $r > 0$ we have

$$\liminf_{n \to \infty} \mathcal{F}(A_n) = \liminf_{n \to \infty} \{\underline{M}(\partial[\![A_n]\!]) - \int_{A_n \cap B_r} h \, d\mathcal{L}^{m+1} - \int_{A_n \cap \complement B_r} h \, d\mathcal{L}^{m+1}\}$$

$$\geq \underline{M}(\partial[\![A]\!]) - \int_{A \cap B_r} h \, d\mathcal{L}^{m+1} - \sup_{n} \int_{A_n \cap \complement B_r} |h| \, d\mathcal{L}^{m+1}.$$

Observing

$$\int_{A \cap B_r} h \, d\mathcal{L}^{m+1} \xrightarrow[r \to \infty]{} \int_A h \, d\mathcal{L}^{m+1},$$

$$\int_{A_n \cap \complement B_r} |h| \, d\mathcal{L}^{m+1} \leq \sup_{n} \mathcal{L}^{m+1}(A_n)^{\frac{m}{m+1}} \cdot \|h\|_{L^{m+1}(G_r)},$$

$G_r := G \cap \complement B_r$, the claim follows.

Since ∂G and $\partial \Omega$ are smooth similar to Theorem 2.3.3 we can prove that $(X \in C_0^1(\mathbb{R}^{m+1}, \mathbb{R}^{m+1}))$

$$\int_{\partial^* A} (\operatorname{div}_{\partial^* A} X + h X \cdot \nu_A) \, d\mathcal{H}^m = \int_{\partial^* A \cap \partial(G-\Omega)} \vartheta \cdot X \cdot \mathcal{N} \{\operatorname{div}_{\partial^* A} \mathcal{N} + h \nu_A \cdot \mathcal{N}\} \, d\mathcal{H}^m$$

for some measurable density $0 \leq \vartheta \leq 1$, moreover $\{\ldots\} \geq 0$ a.e. Here $\mathcal{N}$ is the interior unit normal to $G - \overline{\Omega}$. Now if the curvature conditions relating h and $H_{\partial \Omega}$, $H_{\partial G}$ hold it is easy to check that the right-hand-side of the above equation must vanish so that ∂A has mean curvature h. □

For further details we refer to [35].

Bibliography

[1] W.K. Allard, On the first variation of a varifold. Ann. of Math. 95 (1972), 417–491

[2] F.J. Almgren, Q-valued functions minimizing Dirichlet's integral and the regularity of area minimizing currents up to codimension two. Preprint. Princeton 1983

[3] F.J. Almgren, Optimal isoperimetric inqualities. Indiana Univ. Math. J. 35 (1986) 451–547

[4] J.M. Ball, Convexity conditions and existence theorems in nonlinear elasticity. Arch. Rat. Mech. Anal. Vol. 63 (1977), 337–403

[5] P.G. Ciarlet, Mathematical Elasticity, North-Holland, Amsterdam 1988

[6] J.M. Coron, R. Gulliver, Minimizing p-harmonic maps into spheres. Max-Planck-Institut Preprint No. 13, Bonn 1987

[7] B. Dacorogna, Direct methods in the calculus of variations. Springer Verlag 1989

[8] F. Duzaar, Variational inequalities for harmonic maps. J. Reine Angew. Math. 374 (1987), 39–60

[9] F. Duzaar, M. Fuchs, Optimal regularity theorems for variational problems with obstacles, Manus. Math. 56 (1986), 209–234

[10] F. Duzaar, M. Fuchs, Existence of area minimizing tangent cones of integral currents with prescribed mean curvature. Preprint no. 80 SFB 256, Universität Bonn (1989)

[11] F. Duzaar, M. Fuchs, On removable singularities of p-harmonic maps. Analyse non linéaire, Vol. 7, No. 5 (1990), 385–405

[12] F. Duzaar, M. Fuchs, Existenz und Regularität von Hyperflächen mit vorgeschriebener mittlerer Krümmung. Analysis 10 (1990), 193–230

[13] F. Duzaar, M. Fuchs, On the existence of integral currents with prescribed mean curvature. Manus. Math. 67 (1990), 41–67

[14] F. Duzaar, M. Fuchs, Existence and regularity of functions which minimize certain energies in homotopy classes of mappings. Asymptotic Analysis 5 (1991), 129–144

[15] F. Duzaar, M. Fuchs, On integral currents with constant mean curvature. Rend. Sem. Mat. Univ. Padova, Vol. 85 (1991), 79–103

[16] F. Duzaar, M. Fuchs, Einige Bemerkungen über die Existenz oreintierter Mannigfaltigkeiten mit vorgeschriebener mittlerer Krümmungsform. Zeitschrift Anal. Anw. 10 (1991), 525–534

[17] F. Duzaar, M. Fuchs, A general existence theorem for integral currents with prescribed mean curvature form. Boll. U.M.I. (7) G–B (1991), 901–912

[18] F. Duzaar, M. Fuchs, Einige Bemerkungen über die Regularität von stationären Punkten gewisser geometrischer Variationsintegrale. Math. Nachrichten 152 (1991), 39–47

[19] F. Duzaar, K. Steffen, Boundary regularity for minimizing currents with prescribed mean curvature. Preprint No. 260 SFB 256, Universität Bonn 1992

[20] F. Duzaar, K. Steffen, Area minimizing hypersurfaces with prescribed volume and boundary. Math. Z. 209 (1992), 581–618

[21] J. Eells, J.H. Sampson, Harmonic mappings of Riemannian manifolds. American J. Math. 86 (1964), 109–160

[22] L.C. Evans, Partial regularity for stationary harmonic maps into spheres. Arch. Rat. Mech. Anal. 116 (1991), 101–113

[23] H. Federer, Geometric measure theory. Springer Verlag 1969

[24] H. Federer, W.H. Fleming, Normal and integral currents. Ann. of Math. Vol. 72, 458–520 (1960)

[25] M. Fuchs, Variational inequalities for vector valued functions with non convex obstacles. Analysis 5 (1985), 223–238

[26] M. Fuchs, A note on removable singularities for certain vector valued obstacle problems. Arch. Math. 48 (1987), 521–525

[27] M. Fuchs, Everywhere regularity theorems for mappings which minimize p–energy. Comm. M.U.C. 28, 4 (1987), 673–677

[28] M. Fuchs, The smoothness of the free boundary for a class of vector valued obstacle problems. Comm. P.D.E. 14 (8,9) (1989), 1027–1041

[29] M. Fuchs, Some regularity theorems for mappings which are stationary points of the p–energy functional. Analysis 9 (1989), 127–143

[30] M. Fuchs, p–harmonic obstacle problems. Part I: Partial regularity theory. Ann. Mat. Pura Appl. 156 (1990), 127–158

[31] M. Fuchs, p–harmonic obstacle problems. Part II: Extensions of maps and applications. Manus. Math. 63 (1989), 381–419

[32] M. Fuchs, p–harmonic obstacle problems. Part III: Boundary regularity. Ann. Mat. Pura Appl. 156 (1990), 159–180

[33] M. Fuchs, Hölder continuity of the gradient for degenerate variational inequalities. Nonlinear Analysis, Vol. 15, No. 1 (1990), 85–100

[34] M. Fuchs, On the existence of manifolds with prescribed boundary and mean curvature. TH Darmstadt preprint (1990)

[35] M. Fuchs, Hypersurfaces of prescribed mean curvature enclosing a given body. Manus. Math. 72 (1991), 131–140

[36] M. Fuchs, Smoothness for systems of degenerate variational inequalities with natural growth. Comm. M.U.C., Vol. 33, No. 1 (1992), 33–41

[37] M. Fuchs, Existence via partial regularity for degenerate systems of variational inequalities with natural growth. Comm. M.U.C. Vol. 33, No. 1 (1992), 427–435

[38] M. Fuchs, The blow-up of p-harmonic maps. Manus. Math. 81 (1993), 89–94

[39] M. Fuchs, Regularity for a class of variational integrals motivated by nonlinear elasticity. Asymptotic Analysis 8 (1994), 1–16

[40] M. Fuchs, G. Seregin, Hölder continuity for weak extremals of some two-dimensional variational problems related to nonlinear elasticity. Preprint No. 333 SFB 256, Universität Bonn (1994)

[41] M. Fuchs, G. Seregin, Partial regularity of the deformation gradient for some model problems in nonlinear twodimensional elasticity. Preprint No. 314 SFB 256, Universität Bonn (1993); to appear in St.-Petersburg Math. J.

[42] M. Fuchs, J. Reuling, Partial regularity for certain classes of polyconvex functionals related to nonlinear elasticity. Preprint No. 338 SFB 256, Universität Bonn (1994)

[43] N. Fusco, J. Hutchinson, Partial regularity for minimizers of certain functionals having nonquadratic growth. Ann. Mat. Pura Appl. 156 (1989), 1–24

[44] M. Giaquinta, Multiple integrals in the calculus of variations and non linear elliptic systems. Princeton University Press 1983

[45] M. Giaquinta, E. Giusti, On the regularity of the minima of variational integrals. Acta Math. 148 (1982), 31–40

[46] M. Giaquinta, E. Giusti, The singular set of the minima of certain quadratic functionals. Ann. Sc. Norm. Sup. Pisa, 11 (1984), 45–55

[47] M. Giaquinta, G. Modica, Remarks on the regularity of minimizers of certain degenerate functionals. Manus. Math. 57 (1986), 55–99

[48] E. Giusti, Minimal surfaces and functions of bounded variation. Birkhäuser Monographs in Mathematics 1984

[49] E. Giusti, On the behaviour of the derivatives of minimizers near singular points. Arch. Rat. Mech. Anal. (1986), 137–146

[50] M. Grüter, Eine Bemerkung zur Regularität stationärer Punkte von konform invarianten Variationsproblemen. Manus. Math. 55 (1986), 451–453

[51] R. Gulliver, J. Spruck, Existence theorems for parametric surfaces of prescribed mean curvature. Indiana Univ. Math. J. 22 (1972), 445–472

[52] R. Hardt, D. Kinderlehrer, F.-H. Lin, Existence and regularity of static liquid crystal configurations, IMA preprint series no. 25 (1986)

[53] R. Hardt, F.-H. Lin, Mappings minimizing the L^p–norm of the gradient. Comm. Pure Appl. Math. 11 (1987), 555–588

[54] E. Heinz, Über die Existenz einer Fläche konstanter mittlerer Krümmung bei vorgegebener Berandung. Math. Ann. 127 (1954), 258–287

[55] E. Heinz, On the nonexistence of a surface of constant mean curvature with finite area and prescribed rectifiable boundary. Arch. Rat. Mech. Analysis 35 (1969), 249–252

[56] S. Hildebrandt, Harmonic mappings of Riemannian manifolds. In: Harmonic mappings and minimal immersions. Springer Lecture Notes 1161 (1984), ed. E. Giusti

[57] S. Hildebrandt, Über Flächen konstanter mittlerer Krümmung. Math. Z. 112 (1969), 107–144

[58] S. Hildebrandt, Einige Bemerkungen über Flächen beschränkter mittlerer Krümmung. Math. Z. 115 (1970), 169–178

[59] S. Hildebrandt, Über einen neuen Existenzsatz für Flächen vorgeschriebener mittlerer Krümmung. Math. Z. 119 (1971), 267–272

[60] S. Hildebrandt, J. Jost, K.-O. Widman, Harmonic mappings and minimal surfaces. Inventiones Math. 62 (1980), 269–298

[61] S. Hildebrandt, H. Kaul, K.-O. Widman, An existence theorem for harmonic mappings of Riemannian manifolds. Acta Math. 138 (1977), 1–16

[62] S. Hildebrandt, K.-O. Widman, Variational inequalities for vector valued functions. J. Reine Angew. Math. 309 (1979), 191–220

[63] R. Landes, On the existence of weak solutions of perturbed systems with critical growth. J. Reine Angew. Math. 393 (1989), 21–38

[64] G. Liao, A regularity theorem for harmonic maps with small energy. J. Differential Geom. 22 (1985), 233–241

[65] P. Lindqvist, Regularity for the gradient of the solution to a nonlinear obstacle problem with degenerate ellipticity. Nonlinear Analysis, Vol. 12, No. 2 (1988), 1245–1255

[66] S. Luckhaus, Partial Hölder continuity for minima of certain energies among maps into a Riemannian manifold. Ind. Univ. Math. J. 37 (2) (1988), 346–367

[67] S. Luckhaus, Convergence of minimizers for the p–Dirichlet integral. Preprint no. 129 SFB 256, Universität Bonn (1990)

[68] C.B. Morrey, Multiple integrals in the calculus of variations. Springer Verlag 1966

[69] P. Price, A monotonicity formula for Yang-Mills fields, Manus. Math. 43 (1983), 131–166

[70] Y.G. Reshetnyak, Space mappings with bounded distortion. Translation of Math. Monographs Vol. 73

[71] R. Schoen, K. Uhlenbeck, A regularity theory for harmonic maps. J. Diff. Geom. 17 (1982), 307–335

[72] R. Schoen, K. Uhlenbeck, Boundary regularity and the Dirichlet problem for harmonic maps. J. Diff. Geom. 18 (1983), 253–268

[73] J. Serrin, Local behaviour of solutions of quasi-linear elliptic equations. Acta Math. 111 (1964), 247–302

[74] L. Simon, Lectures on geometric measure theory. Proc. C.M.A. Australian National University, Vol. 3 (1983)

[75] J. Simons, Minimal varieties in Riemannian manifolds. Ann. of Math. 88 (1968), 62–105

[76] K. Steffen, On the existence of surfaces with prescribed mean curvature and boundary. Math. Z. 146 (1976), 113–135

[77] E.M. Stein, Singular integrals and differentiability properties of functions. Princeton University Press 1970

[78] K. Steffen, Isoperimetric inequalities and the problem of Plateau. Math. Annalen 222 (1976), 97–144

[79] V. Šverák, Regularity properties of deformations with finite energy. Arch. Rat. Mech. Anal. Vol. 98 (1987), 105–127

[80] K. Uhlenbeck, Regularity for a class of nonlinear elliptic systems. Acta Math. 138 (1977), 219–240

[81] H. Wente, An existence theorem for surfaces of constant mean curvature. J. Math. Analysis Appl. 26 (1969), 318–344

Index

asymptotically regular 58, 69
blow-up limit 36
boundary regularity 4, 38
Caccioppoli set 135
Caccioppoli inequality 6, 17, 23, 32, 43
codimension one 51, 52, 118
compactness theorem 103, 138
constant mean curvature 95, 119, 123
contact set 5, 63
current 52, 99, 101
degenerate functional 5
dimension reduction 31, 119, 130
Euler equation 6, 9, 15, 60
extension of maps 17
finite perimeter 135
first variation formula 107
geodesic ball 19, 41, 50
H-surface 51, 88
harmonic map 5, 40, 45, 48
Hausdorff dimension 3
Hausdorff measure 3
homotopy problem 48, 49
hybrid inequality 22, 23
integer multiplicity current 101
isolated singularity 45, 55, 118
isoperimetric inequality 52, 78, 103, 104
isoperimetric problem 123, 125, 130
maximum principle 7, 41, 58
max. principle for the mass 112, 120
mean curvature form 90
mean curvature problem 84, 105
mean curvature vector 85
monotonicity formula 25, 43, 119
nonlinear elasticity 5, 70, 74
obstacle problem 1, 5, 6, 7, 41
oriented volume 106
p-harmonic map 5, 40
p-stationary map 40
partial higher integrability 26
principal curvatures 114
rectifiable set 102
reduced boundary 136

regular ball 41, 42
regular set 2, 117
relaxation 70
removable singularity 45
reverse Hölder inequality 29
singular set 2
slice (of a current) 120
small range condition 42, 46
small solution 134
tangent cone 119, 128
tangent map 32, 36
variation 11, 13, 76, 113
variational inequality 42, 58, 63
varifold 105, 117

Lattices and Codes

A Course Partially Based on Lectures by F. Hirzebruch

by Wolfgang Ebeling

1994. xvi, 178 pages. (Advanced Lectures in Mathematics; edited by Martin Aigner, Gerd Fischer, Michael Grüter, Manfred Knebusch, and Gisbert Wüstholz) Softcover
ISBN 3-528-06497-8

From the contents: Lattices and Codes – Theta Functions and Weight Enumerators – Even Unimodular Lattices – The Leech Lattice – Lattices over Integers of Number Fields and Self-Dual Codes

The purpose of coding theory is the design of efficient systems for the transmission of information. The mathematical treatment leads to certain finite structures: the error-correcting codes. Surprisingly problems which are interesting for the design of codes turn out to be closely related to problems studied partly earlier and independently in pure mathematics.

In this book, examples of such connections are presented. The relation between lattices studied in number theory and geometry and error-correcting codes is discussed. The book provides at the same time an introduction to the theory of integral lattices and modular forms and to coding theory.

Vieweg Publishing · P.O. Box 58 29 · D-65048 Wiesbaden

Lectures on Nonlinear Evolution Equations

Initial Value Problems

by Reinhard Racke

1992. viii, 259 pages (Aspects of Mathematics, Vol. E19; edited by Klas Diederich) Hardcover
ISBN 3-528-06421-8

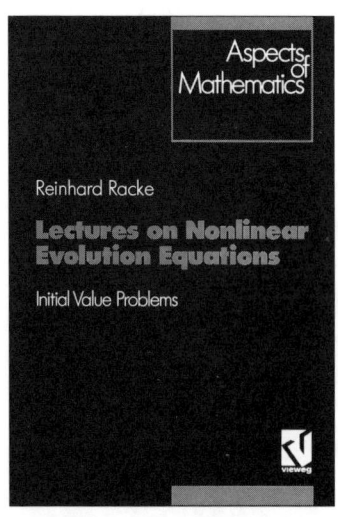

This book serves as an elementary, self contained introduction into some important aspects of the theory of global solutions to initial value problems for nonlinear evolution equations. The presentation is made using the classical method of continuation of local solutions with the help of a priori estimates obtained for small data. The existence and uniqueness of small, smooth solutions which are defined for all values of the time parameter is investigated. Moreover, the asymptotic behaviour of the solutions is described as time tends to infinity. The methods are discussed in detail for nonlinear wave equations. Other examples are the equations of elasticity, heat equations of thermoelasticity, Schrödinger equations, Klein-Gordon equations, Maxwell equations and plate equations. To emphasize the necessity of studying conditions under which small data problems allow global solutions some blow-up results are shortly described. Moreover, a prospect on corresponding initial boundary value problems and on open questions is given.

Vieweg Publishing · P.O. Box 58 29 · D-65048 Wiesbaden